Springer Aerospace Technology

The series explores the technology and the science related to the aircraft and spacecraft including concept, design, assembly, control and maintenance. The topics cover aircraft, missiles, space vehicles, aircraft engines and propulsion units. The volumes of the series present the fundamentals, the applications and the advances in all the fields related to aerospace engineering, including:

- structural analysis,
- aerodynamics,
- aeroelasticity,
- aeroacoustics,
- flight mechanics and dynamics
- orbital maneuvers,
- avionics,
- systems design,
- materials technology,
- launch technology,
- payload and satellite technology,
- space industry, medicine and biology.

The series' scope includes monographs, professional books, advanced textbooks, as well as selected contributions from specialized conferences and workshops.

The volumes of the series are single-blind peer-reviewed.

To submit a proposal or request further information, please contact: Mr. Pierpaolo Riva at pierpaolo.riva@springer.com (Europe and Americas) Mr. Mengchu Huang at mengchu.huang@springer.com (China)

The series is indexed in Scopus and Compendex

Lianda Wu · Mingjiang Zhang

Hansen Coefficients in Satellite Orbital Dynamics

南京大学出版社
NANJING UNIVERSITY PRESS

Springer

Lianda Wu
Purple Mountain Observatory
Chinese Academy of Sciences
Nanjing, Jiangsu, China

Mingjiang Zhang
Purple Mountain Observatory
Chinese Academy of Sciences
Nanjing, Jiangsu, China

ISSN 1869-1730 ISSN 1869-1749 (electronic)
Springer Aerospace Technology
ISBN 978-981-97-0455-2 ISBN 978-981-97-0456-9 (eBook)
https://doi.org/10.1007/978-981-97-0456-9

Jointly published with Nanjing University Press
The print edition is not for sale in China (Mainland). Customers from China (Mainland) please order the print book from: Nanjing University Press.

This Springer imprint is published by the registered company Springer Nature Singapore Pte Ltd.
The registered company address is: 152 Beach Road, #21-01/04 Gateway East, Singapore 189721, Singapore

Contents

Chapter 1
Introduction

1.1 Introduction of Hansen Coefficients

Disturbing functions need to be expanded as explicit functions of time and orbital elements in analytical theories of celestial mechanics, which involve two types of commonly used special functions, namely inclination functions and Hansen coefficients. Refer to [1] for the detailed calculation of inclination functions.

In satellite dynamics, two types of expansion of disturbing functions are related to Hansen coefficients.

1.1.1 Disturbing Function of the Earth's Gravitational Field

The disturbing function of the Earth's gravitational field is

$$R = \sum_{l=2}^{\infty}\sum_{m=0}^{l}\frac{\mu R_E^l}{r^{l+1}}\overline{P}_l^m(\sin\varphi)\left(\overline{C}_{lm}\cos m\lambda + \overline{S}_{lm}\sin m\lambda\right) \tag{1.1}$$

where μ is the geocentric gravitational constant, R_E is the mean equatorial radius of the Earth, r is the geocentric distance of the satellite, λ and φ are the longitude and latitude of the satellite respectively, $\overline{P}_l^m(\sin\varphi)$ is the normalized associated Legendre polynomial, and $\overline{C}_{lm}$ and $\overline{S}_{lm}$ are the normalized coefficients of the Earth's gravitational field.

The expansion of the disturbing function of the Earth's gravitational field in Eq. (1.1) with the aid of the inclination function and Hansen coefficient is [2, 3]

$$R = \sum_{lmpq}\frac{\mu R_E^l}{a^{l+1}}J_{lm}\overline{F}_{lmp}(I)X_{l-2p+q}^{-l-1,l-2p}(e)\cos(\psi + \lambda_{lm}) \tag{1.2}$$

L.D Wu and M.J Zhang, *Hansen Coefficients in Satellite Orbital Dynamics*,
Springer Aerospace Technology, https://doi.org/10.1007/978-981-97-0456-9_1

where $J_{lm} = \left(\overline{C}_{lm}^2 + \overline{S}_{lm}^2\right)^{1/2}$; $\psi = (l-2p)\omega + (l-2p+q)M + m(\Omega - \theta)$; a, e, I, Ω, ω and M are the Keplerian elements of the satellite; θ is the Greenwich sidereal time, $\lambda_{lm} = \frac{l-m}{2}\pi - \arctan\left(\frac{\overline{S}_{lm}}{\overline{C}_{lm}}\right)$; $\overline{F}_{lmp}(I)$ is the normalized inclination function [2], and $X_{l-2p+q}^{-l-1,l-2p}(e)$ is the Hansen coefficient.

1.1.2 Disturbing Function of Luni-Solar Perturbations

The disturbing function of luni-solar perturbations is

$$R = \frac{Gm'}{r'}\sum_{l=2}^{\infty}\frac{1}{2l+1}\left(\frac{r}{r'}\right)^l \sum_{m=0}^{l}\overline{P}_l^m(\sin\varphi)\overline{P}_l^m(\sin\varphi')\cos m(\lambda - \lambda') \tag{1.3}$$

where G is the gravitational constant, m' is the mass of the perturbing body, r is the geocentric distance of the satellite, λ and φ are the longitude and latitude of the satellite respectively, r' is the geocentric distance of the perturbing body, λ' and φ' are the longitude and latitude of the perturbing body respectively, and $\overline{P}_l^m(\sin\varphi)$ and $\overline{P}_l^m(\sin\varphi')$ are the normalized associated Legendre polynomials.

In general, only the second-order term needs to be considered for solar perturbation, whereas the second-order and third-order terms need to be considered for lunar perturbation. The expansion of the disturbing function for luni-solar perturbations Eq. (1.3) with the aid of the inclination functions and Hansen coefficients is [2, 3]:

$$R = \sum_{lmpp'qq'}\frac{Gm'(-1)^{l+m}a^l}{(2l+1)a'^{l+1}}\overline{F}_{lmp}(I)\overline{F}_{l,m,l-p'}(I')X_{l-2p+q}^{l,l-2p}(e)X_{l-2p'+q'}^{-l-1,l-2p'}(e')\cos\psi_1 \tag{1.4}$$

where a, e, I, Ω, ω and M are the Keplerian elements of the satellite; a', e', I', Ω', ω' and M' are the Keplerian elements of the perturbing body; $\overline{F}_{lmp}(I)$ and $\overline{F}_{l,m,l-p'}(I')$ are the normalized inclination functions [2]; $X_{l-2p+q}^{l,l-2p}(e)$ and $X_{l-2p'+q'}^{-l-1,l-2p'}(e')$ are the Hansen coefficients; and $\psi_1 = (l-2p)\omega + (l-2p+q)M + (l-2p')\omega' + (l-2p'+q')M' + m(\Omega - \Omega')$.

There are thus two types of Hansen coefficients, namely $X_k^{-l-1,l-2p}$ and $X_k^{l,l-2p}$, in satellite dynamics. They are defined by

$$\left(\frac{a}{r}\right)^{l+1}\exp[j(l-2p)f] = \sum_{k=-\infty}^{\infty}X_k^{-l-1,l-2p}\exp(jkM) \tag{1.5}$$

$$\left(\frac{r}{a}\right)^{l}\exp[j(l-2p)f] = \sum_{k=-\infty}^{\infty}X_k^{l,l-2p}\exp(jkM) \tag{1.6}$$

where $j = \sqrt{-1}$ is the imaginary unit and f is the true anomaly.

Kaula referred to these Hansen coefficients as eccentricity functions [4–6] expressed as

$$\begin{aligned} G_{lpq}(e) &= X_{l-2p+q}^{-l-1,l-2p}(e) \\ H_{lpq}(e) &= X_{l-2p+q}^{l,l-2p}(e) \end{aligned} \tag{1.7}$$

1.2 Definition of the Hansen Coefficient and its Relation with Classical Coefficients

The Hansen coefficient $X_k^{n,m}$ is defined as [7]

$$\left(\frac{r}{a}\right)^n \exp(jmf) = \sum_{k=-\infty}^{\infty} X_k^{n,m}(e)\exp(jkM) \tag{1.8}$$

where the function $\left(\frac{r}{a}\right)^n \exp(jmf)$ is called the generating function of the Hansen coefficient. This generating function is important to the expansion of the disturbing function, and its expansion has thus received much attention in the field of celestial mechanics.

Cayley presented the expansions [8–10]

$$\left(\frac{r}{a}\right)^n \cos mf = \sum_{k=0}^{\infty} C_k^{n,m}(e)\cos kM \tag{1.9}$$

$$\left(\frac{r}{a}\right)^n \sin mf = \sum_{k=0}^{\infty} S_k^{n,m}(e)\sin kM \tag{1.10}$$

as well as tables of the coefficients $C_k^{n,m}(e)$ and $S_k^{n,m}(e)$ ($n = -5, -4, \cdots, -1, 1, 2, \cdots, 4; 0 < m < |n|$) up to the seventh order in eccentricity, e^7. Evidently, the relations between the Hansen coefficient $X_k^{n,m}(e)$ and $C_k^{n,m}(e)$ and $S_k^{n,m}(e)$ are [9, 10]

$$\begin{aligned} C_0^{n,m} &= X_0^{n,m} \\ C_k^{n,m} &= X_k^{n,m} + X_{-k}^{n,m}, \quad X_k^{n,m} = \frac{C_k^{n,m} + S_k^{n,m}}{2} \\ S_k^{n,m} &= X_k^{n,m} - X_{-k}^{n,m}, \quad X_{-k}^{n,m} = \frac{C_k^{n,m} - S_k^{n,m}}{2} \end{aligned} \tag{1.11}$$

Newcomb presented the expansion [10, 11]

$$\left(\frac{r}{a}\right)^n \exp[jm(f-M)] = \sum_{k=-\infty}^{\infty} N_k^{n,m} \exp(jkM) \tag{1.12}$$

where

$$N_k^{n,m} = \sum_{q=0}^{\infty} \Pi_k^q(n,m)e^q, \quad |q-k| = 0, 2, 4, \cdots \tag{1.13}$$

The factor $\Pi_k^q(n,m)$ is called the Newcomb operator. The relation between the Hansen coefficient and $N_k^{n,m}$ is [10]

$$N_k^{n,m} = X_{m+k}^{n,m} \tag{1.14}$$

Newcomb presented tables of $N_k^{n,m}$ truncated to the eighth order in eccentricity, e^8 [11].

Iszak et al. [12] expressed the Newcomb operator in the form

$$X_{\rho,\sigma}^{n,m} = \Pi_{\rho-\sigma}^{\rho+\sigma}(n,m) \tag{1.15}$$

Through variable transformation, the Hansen coefficient is expressed as [10]

$$X_k^{n,m} = e^{|m-k|} \sum_{\sigma=0}^{\infty} X_{\sigma+\alpha,\sigma+\beta}^{n,m} e^{2\sigma} \tag{1.16}$$

where $\alpha = \max(0, k-m)$ and $\beta = \max(0, m-k)$. Iszak et al. expanded the Hansen coefficients up to the eighth order in eccentricity, e^8 [10, 12], and Cherniack further expanded the coefficients up to the twelfth order in eccentricity, e^{12}[9, 10].

The results [8, 9, 11, 12] presented by Cayley, Newcomb, Izsak et al. and Cherniack are obviously useful for the calculation of Hansen coefficients, but their truncation orders in eccentricity are too low. We thus need to study the calculation of Hansen coefficients further.

1.3 Requirements of Hansen Coefficients in Satellite Dynamics

Generally, a meter-level accuracy of perturbation calculations is required in the analytical theory of satellite dynamics. If the accuracy threshold is set at 1 meter, it is enough to take the terms with $l = 2$ and 3 into account in the expansion of the disturbing function of luni-solar perturbations. However, because the coefficients of the Earth's gravitational field converge slowly, many terms need to be considered in the expansion of the disturbing function of the Earth's gravitational field; i.e.,

Table 1.1 Truncation orders N of the Earth's gravitational field for the precise orbit determination of laser satellites at different altitudes [13]

Orbital semi-major axis a (km)	Satellite	Truncation order N
7331	Starlette	90
12,270	LAGEOS	20
26,600	GPS	12

$$\Delta\sigma = \sum_{l=2}^{N}\sum_{m=0}^{l}\sum_{p=0}^{l}\sum_{q=Q_1}^{Q_2} \Delta\sigma_{lmpq} \tag{1.17}$$

where N is the truncation order of the Earth's gravitational field and $[Q_1, Q_2]$ is the summation range of the index q. The perturbation of the Earth's gravitational field can be divided into terms having different magnitudes.

(1) Terms with $m = 0$ arise from zonal harmonics and must be taken into account because of their secular and long-period perturbations.
(2) Terms with $k = l - 2p + q = 0$ are m-daily terms whose magnitudes are sufficiently large that they need to be taken into account generally.
(3) Other terms are short-period perturbations arising from tesseral harmonics and relate not only to the coefficients $J_{lm}(m = 1, 2, \cdots, l)$ of the Earth's gravitational field but also to the quantity a^{-l} associated with the semi-major axis a, the inclination function $\overline{F}_{lmp}(I)$ and the eccentricity function $G_{lpq}(e)$.

To calculate the perturbation of the Earth's gravitational field well, the truncation order N must first be determined. Table 1.1 presents the truncation orders N recommended by the International Earth Rotation Service (IERS) [13] for the precise orbit determination of laser satellites.

However, orbits with large eccentricities are not considered in Table 1.1. The minimum values of the semi-major axes $a_{\min}$ for orbits with different eccentricities e can be approximately estimated using the formula

$$a_{\min} = \frac{6378.14 + 200}{6378.14(1 - e)} \tag{1.18}$$

Note that in Eq. (1.18), the unit of calculation for $a_{\min}$ is the Earth's equatorial radius R_E. The minimum values of semi-major axes $a_{\min}$ corresponding to orbits with different eccentricities e are presented in Table 1.2.

Besides the orbital semi-major axis a, the truncation order N of the Earth's gravitational field also relates to the inclination function $\overline{F}_{lmp}(I)$ and the eccentricity function $G_{lpq}(e)$. When the orbital eccentricity is large, the values of Hansen coefficients or eccentricity functions are large. Therefore, the truncation order N of the Earth's gravitational field can only be determined after evaluating the Hansen coefficients, which is the primary purpose of our study of Hansen coefficients.

Table 1.2 Minimum values of semi-major axes $a_{\min}$ corresponding to orbits with different eccentricities e

Eccentricity e	$a_{\min}$(km)	$a_{\min}(R_E)$
0.1	7309	1.145952
0.2	8223	1.289196
0.3	9397	1.473367
0.4	10,964	1.718929
0.5	13,156	2.062714
0.6	16,445	2.578393
0.7	21,927	3.437857
0.75	26,313	4.125428

When calculating the perturbation of the Earth's gravitational field, the summation range of the index q in Eq. (1.17) is generally from $-Q$ to $+Q$ (where Q is the maximum of the index q); i.e., the magnitude of the term with the index $q = 0$ is deemed to be the largest. However, Wnuk presented the relationships of the eccentricity function G_{lpq} and its derivative $\frac{dG_{lpq}}{de}$ versus the index q for the given indices $l = 66$ and $p = 20$ and given eccentricity $e = 0.4$ [14]; see Fig. 1.1. The figure shows that the eccentricity function G_{lpq} and its derivative $\frac{dG_{lpq}}{de}$ ($l = 66$, $p = 20$) have maximum values at $q = 39$ and $q = 41$ respectively, when the eccentricity $e = 0.4$. In addition, Wnuk presented the relationship of the derivative of the Hansen coefficient (or eccentricity function) $\frac{dG_{lpq}}{de}$ versus the index q for the given indices $l = 40$ and $p = 14$ and different eccentricities [14]; see Fig. 1.2. The figure shows that the derivative of the Hansen coefficient (or eccentricity function) is not a maximum at $q = 0$ when the eccentricity $e = 0.1, 0.2, \cdots, 0.8$. Thus, to calculate the perturbation well, the Hansen coefficients need to be studied, and the calculation of Hansen coefficients for each index l is required, especially for orbits having large eccentricity.

As seen in the following chapters, there are difficulties in calculating Hansen coefficients. Some direct calculation methods can be unstable whereas recursion calculation methods are not efficient. It is thus necessary to study the characteristics and calculation methods of Hansen coefficients in the analytical theory of satellite dynamics.

1.4 Overview of the Present Monograph

This monograph comprises five chapters and four appendices.

In this chapter, the Hansen coefficient is defined in terms of the expansion of disturbing functions. According to the magnitudes of perturbation terms, the terms of zonal harmonics, the m-daily terms and the short-period terms arising from tesseral harmonics in the perturbation of the Earth's gravitational field are calculated separately; and the truncation order N of the perturbation of the Earth's gravitational field

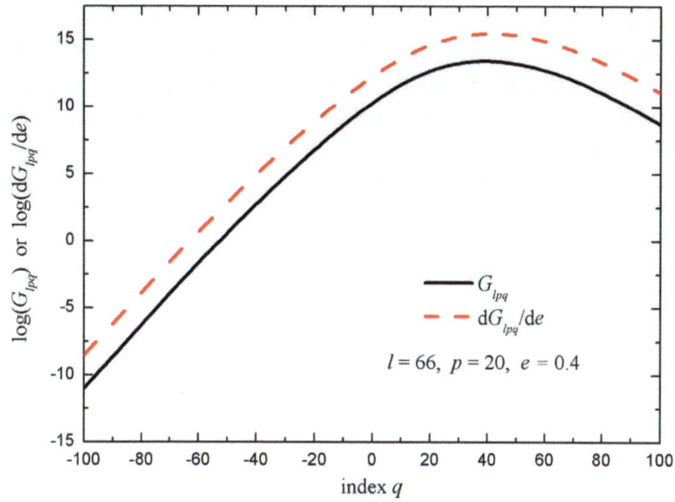

Fig. 1.1 Eccentricity function G_{lpq} and its derivative $\frac{dG_{lpq}}{de}$ versus the index q ($e = 0.4, l = 66, p = 20$)[14]

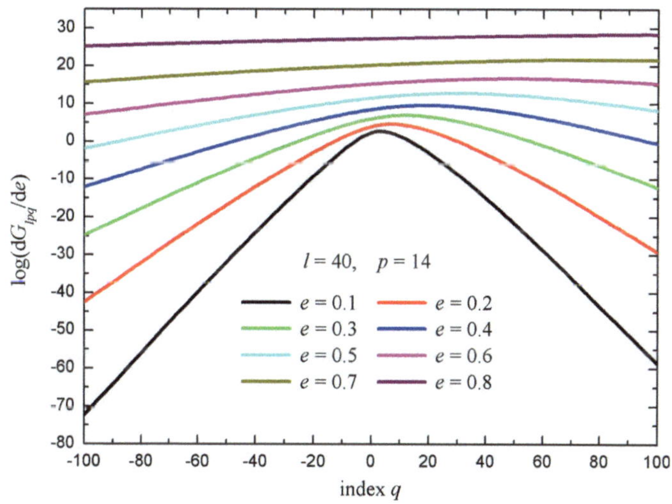

Fig. 1.2 Relationship of the derivative of the Hansen coefficient (or eccentricity function) $\frac{dG_{lpq}}{de}$ versus the index q for different eccentricities ($l = 40, p = 14$) [14]

and the summation range of the index q relate to the magnitudes and characteristics of the Hansen coefficients.

In Chap. 2, the calculation of Hansen coefficients $X_0^{n,m}$ and $X_0^{-(n+1),m}$ with $k = 0$ and their derivatives is studied. The corresponding perturbations are m-daily terms. Direct and recursion methods of calculating the Hansen coefficients $X_0^{n,m}$ and

$X_0^{-(n+1),m}$ and their derivatives are presented, two recursion formulae are deduced using the hypergeometric series [15], and the sixteen methods are compared. All these methods provide the required calculation accuracy, and the recursion methods have high computational efficiencies. The methods used to calculate Hansen coefficients for the perturbations only satisfy the requirement to calculate perturbations without singularities; i.e., the calculation of Hansen coefficients and their derivatives for m-daily perturbation terms has been satisfactorily settled.

The calculation of Hansen coefficients with $k \neq 0$ and their derivatives is studied in Chaps. 3, 4 and 5. Direct calculation methods are discussed in Chap. 3, recursion calculation methods in Chap. 4 and other methods in Chap. 5.

In Chap. 3, several direct methods of calculating Hansen coefficients and their derivatives are reviewed. By directly differentiating the mathematical expressions of the relevant methods with respect to the orbital eccentricity e, the direct calculation expressions of the derivatives of Hansen coefficients are derived. The computational efficiencies of these methods are compared, and their computational stabilities are analyzed. In calculation examples, it is shown that [16]

(1) for small orbital eccentricities $e < 0.2$, all the calculation methods satisfy the requirement of calculation accuracy;
(2) for large orbital eccentricities, relatively satisfactory calculation results can be obtained using Wnuk's method [14] and Giacaglia's method [17];
(3) McClain's first method [18, 19] is suitable for perturbation calculations without singularities;
(4) the main difficulty encountered in the calculation of Hansen coefficients is the insufficient word length of the computer; and
(5) a criterion is required to determine the stabilities of the calculation methods.

Wnuk's method [14] is not applicable to perturbation calculations without singularities. Therefore, in calculating Hansen coefficients with $k \neq 0$ and their derivatives, it is reasonable to adopt Wnuk's method [14] for large orbital eccentricities and McClain's first method [18, 19] for small orbital eccentricities, in the case of double-precision computation; McClain's first method [18, 19] with quadruple precision computation covers the calculations of Hansen coefficients for both large and small orbital eccentricities, and it is thus a recommended choice.

In Chap. 4, the recursion formulae of Hansen coefficients given by Giacaglia [17] and McClain [18] are reviewed. The analysis reveals that there are only five independent recursion formulae among those given by Giacaglia and McClain, which are called the basic recursion formulae. A remarkable characteristic of the recursion formulae of Hansen coefficients—the subscripts k of Hansen coefficients $X_k^{-(n+1),m}$ and $X_k^{n,m}$ in the recursion formulae are the same—is analyzed. On the basis of the systematic classification of recursions of Hansen coefficients, the deficiencies of existing basic recursion formulae are pointed out, and two new recursion formulae of Hansen coefficients are deduced [20] as an effective supplement of basic recursion formulae. Recursion methods of ordinary Hansen coefficients and eccentricity functions are investigated, the computational efficiencies of direct and recursion calculation methods are compared, and a type of batch recursion method with high efficiency is presented [21]. Finally, it is shown that the recursion relations of Hansen coefficients can be used to determine the stabilities of direct calculation methods [16], and the calculation of Hansen coefficients in perturbation calculations for large orbital eccentricities is discussed.

In Chap. 5, other methods of calculating Hansen coefficients, including harmonic analysis, Taylor series expansion, interpolation calculation and Chebyshev polynomial approximation, are introduced, and the applicability and accuracy of the corresponding methods are discussed.

The appendices present the deduction of important formulae and the Fortran programs of the main calculation methods of Hansen coefficients.

References

1. Wu LD, Wang HB, Ma JY (2012) Inclination function in satellite dynamics. Science Press, Beijing
2. Gaposchkin EM (1973) 1973 Smithsonian standard earth (III). SAO Special Report No. 353, Smithsonian Astrophysical Observatory, Cambridge, Massachusetts
3. Wu LD (2011) Orbit and detection of artificial satellites and space debris. China Science and Technology Press, Beijing (In Chinese)
4. Kaula WM (1966) Theory of satellite geodesy: applications of satellites to geodesy. Blaisdell Publishing Company
5. Kaula WM (1961) Analysis of gravitational and geometric aspects of geodetic utilization of satellites. Geophys J Int 5(2):104–133
6. Kaula WM (1962) Development of the lunar and solar disturbing function for a close satellite. Astron J 67(5):300–303
7. Plummer HC (1918) An introductory treatise on dynamical astronomy. Cambridge University Press, London
8. Cayley A (1861) Tables of the developments of functions in the theory of elliptic motion. Mem R Astron Soc 29:191–306
9. Cherniack JR (1972) Computation of Hansen coefficients. SAO Special Report No. 346, Smithsonian Astrophysical Observatory
10. Hughes S (1981) The computation of tables of Hansen coefficients. Celest Mech 25(1):101–107
11. Newcomb S (1895) A development of the perturbative function in cosines of multiples of the mean anomalies and of angles of multiples of the mean anomalies and of angles between the perihelia and common node and in powers of the eccentricities and mutual inclination. U.S. Nautical Almanac Office, Washington
12. Izsak IG, Gerard JM, Efimba R, Barnett MP (1964) Construction of Newcomb operators on a digital computer. SAO Special Report No. 140, Smithsonian Astrophysical Observatory
13. Petit G, Luzum B (eds) (2010) IERS conventions (2010), IERS Technical Note No. 36, Verlag des Bundesamts für Kartographie und Geodäsie, Frankfurt am Main
14. Wnuk E (1997) Highly eccentric satellite orbits. Adv Space Res 19(11):1735–1740
15. Wu LD, Zhang MJ (2012) Recursive methods for computation of Hansen coefficients $X_0^{n,m}$, $X_0^{-(n+1),m}$ and their derivatives. J Astrodyn 2(4):1–10 (In Chinese)
16. Wu LD, Zhang MJ (2022) Direct calculation methods of Hansen coefficients and their derivatives. Chin Astron Astrophy 46(1):137–151
17. Giacaglia GEO (1976) A note on Hansen's coefficients in satellite theory. Celest Mech 14(4):515–523
18. McClain WD (1978) A recursively formulated first-order semianalytic artificial satellite theory based on the generalized method of averaging. Volume II. The explicit development of the

first-order averaged equations of motion for the nonspherical gravitational and nonresonant third-body perturbations. NASA CR-156783
19. Proulx RJ, McClain WD (1988) Series representations and rational approximations for Hansen coefficients. J Guidance 11(4):313–319
20. Wu LD, Zhang MJ (2021) Two new recursion formulae of Hansen coefficients. Chin Astron Astrophy 45(4):531–541
21. Wu LD, Zhang MJ (2022) Computational efficiency of the recursion of Hansen coefficients. Chin Astron Astrophy 46(2):65–72

Chapter 2
Calculation of $X_0^{n,m}$ and $X_0^{-(n+1),m}$ and Their Derivatives

The Hansen coefficient $X_k^{n,m}$ is defined as [1]

$$\left(\frac{r}{a}\right)^n \exp(jmf) = \sum_{k=-\infty}^{\infty} X_k^{n,m}(e)\exp(jkM) \tag{2.1}$$

where r is the geocentric distance, a is the semi-major axis, e is the eccentricity, M is the mean anomaly, f is the true anomaly, and $j = \sqrt{-1}$. According to the orthogonality of trigonometric functions, the Hansen coefficient $X_k^{n,m}(e)$ is given by the integral

$$X_k^{n,m}(e) = \frac{1}{2\pi}\int_0^{2\pi}\left(\frac{r}{a}\right)^n \cos(mf - kM)dM \tag{2.2}$$

When $k = 0$, Eq. (2.2) becomes

$$X_0^{n,m}(e) = \frac{1}{2\pi}\int_0^{2\pi}\left(\frac{r}{a}\right)^n \cos mf dM \tag{2.3}$$

Therefore, the Hansen coefficient $X_0^{n,m}$ is the average value of $\left(\frac{r}{a}\right)^n \cos mf$:

$$X_0^{n,m}(e) = \left\langle\left(\frac{r}{a}\right)^n \cos mf\right\rangle \tag{2.4}$$

where $\langle x\rangle$ refers to the average value of x.

L.D Wu and M.J Zhang, *Hansen Coefficients in Satellite Orbital Dynamics*, Springer Aerospace Technology, https://doi.org/10.1007/978-981-97-0456-9_2

2.1 Series Expressions of $X_0^{n,m}$ and $X_0^{-(n+1),m}$

Laskar and Boué presented the series expression of the Hansen coefficient $X_0^{n,m}$ (see Appendix A.1 for the specific deduction) [2]:

$$X_0^{n,m} = (-1)^m \frac{(n+1+m)!}{(n+1)!} \sum_{l=0}^{[(n+1-m)/2]} \frac{(n+1-m)!}{l!(m+l)!(n+1-m-2l)!} \left(\frac{e}{2}\right)^{m+2l} \tag{2.5}$$

The expressions of average values for the common functions related to $X_0^{n,m}$ are

$$X_0^{0,m} = \langle \cos mf \rangle = \left(1+\sqrt{1-e^2}\right)\left(\frac{-e}{1+\sqrt{1-e^2}}\right)^m \tag{2.6}$$

$$X_0^{-1,m} = \left\langle \left(\frac{a}{r}\right) \cos mf \right\rangle = \left(\frac{-e}{1+\sqrt{1-e^2}}\right)^m \tag{2.7}$$

$$X_0^{-(n+1),m} = \left\langle \left(\frac{a}{r}\right)^{n+1} \cos mf \right\rangle = \left(1-e^2\right)^{-(n-1/2)} \sum_{k=0}^{[(n-1-m)/2]} \binom{n-1}{m+2k}\binom{m+2k}{k}\left(\frac{e}{2}\right)^{m+2k} \tag{2.8}$$

where n and m are positive integers, e is the eccentricity, and $[x]$ is the integral part of x.

According to the above expressions, both $X_0^{n,m}$ and $X_0^{-(n+1),m}$ have the factor $e^{|m|}$; i.e.,

$$X_0^{n,m} \sim O\left(e^{|m|}\right), \quad X_0^{-(n+1),m} \sim O\left(e^{|m|}\right) \tag{2.9}$$

To give readers a visual concept of the Hansen coefficients $X_0^{n,m}$ and $X_0^{-n,m}$, the specific expressions of $X_0^{n,m}$ of order 0–8 ($0 \le n \le 8$, $0 \le m \le n$) are listed in Table 2.1, and the specific expressions of $X_0^{-n,m}$ of order 2–10 ($2 \le n \le 10, 0 \le m \le n-2$) are listed in Table 2.2. The corresponding eccentricity functions of Hansen coefficients $X_0^{n,m}$ and $X_0^{-n,m}$, $H_{n,(n-m)/2,-m} = X_0^{n,m}$ and $G_{n-1,(n-1-m)/2,-m} = X_0^{-n,m}$, are presented in the remarks of Tables 2.1 and 2.2 respectively. Refer to [2, 3] for the higher-order expressions.

For example, when $n = 7$ and $m = 2$, the look-up Table 2.1 is as follows.

n	m	Power exponent	Coefficients					Remark
		e	e^0	e^2	e^4	e^6	e^8	
7	2	2	45/4	225/8	675/64	45/128		

Therefore,

Table 2.1 Series expressions of $X_0^{n,m}$ $(0 \le n \le 8, 0 \le m \le n)$

n	m	Power exponent	Coefficients					Remark
		e	e^0	e^2	e^4	e^6	e^8	
0	0	0	1					H_{000}
1	0	0	1	1/2				
1	1	1	−3/2					H_{10-1}, H_{111}
2	0	0	1	3/2				H_{210}
2	1	1	−2	−1/2				
2	2	2	5/2					H_{20-2}, H_{222}
3	0	0	1	3	3/8			
3	1	1	−5/2	−15/8				H_{31-1}, H_{321}
3	2	2	15/4	5/8				
3	3	3	−35/8					H_{30-3}, H_{333}
4	0	0	1	5	15/8			H_{420}
4	1	1	−3	−9/2	−3/8			
4	2	2	21/4	21/8				H_{41-2}, H_{432}
4	3	3	−7	−7/8				
4	4	4	63/8					H_{40-4}, H_{444}
5	0	0	1	15/2	45/8	5/16		
5	1	1	−7/2	−35/4	−35/16			H_{52-1}, H_{531}
5	2	2	7	7	7/16			
5	3	3	−21/2	−63/16				H_{51-3}, H_{543}
5	4	4	105/8	21/16				
5	5	5	−231/16					H_{50-5}, H_{555}
6	0	0	1	21/2	105/8	35/16		H_{630}
6	1	1	−4	−15	−15/2	−5/16		
6	2	2	9	15	45/16			H_{62-2}, H_{642}
6	3	3	−15	−45/4	−9/16			
6	4	4	165/8	99/16				H_{61-4}, H_{654}
6	5	5	−99/4	−33/16				
6	6	6	429/16					H_{60-6}, H_{666}
7	0	0	1	14	105/4	35/4	35/128	
7	1	1	−9/2	−189/8	−315/16	−315/128		H_{73-1}, H_{741}
7	2	2	45/4	225/8	675/64	45/128		
7	3	3	−165/8	−825/32	−495/128			H_{72-3}, H_{753}
7	4	4	495/16	297/16	99/128			

(continued)

Table 2.1 (continued)

n	m	Power exponent	Coefficients					Remark
		e	e^0	e^2	e^4	e^6	e^8	
7	5	5	−1287/32	−1287/128				H_{71-5}, H_{765}
7	6	6	3003/64	429/128				
7	7	7	−6435/128					H_{70-7}, H_{777}
8	0	0	1	18	189/4	105/4	315/128	H_{840}
8	1	1	−5	−35	−175/4	−175/16	-35/128	
8	2	2	55/4	385/8	1925/64	385/128		H_{83-2}, H_{852}
8	3	3	−55/2	−825/16	−495/32	−55/128		
8	4	4	715/16	715/16	715/128			H_{82-4}, H_{864}
8	5	5	−1001/16	−1001/32	−143/128			
8	6	6	5005/64	2145/128				H_{81-6}, H_{876}
8	7	7	−715/8	−715/128				
8	8	8	12,155/128					H_{80-8}, H_{888}

$$X_0^{7,2} = e^2\left(\frac{45}{4} + \frac{225}{8}e^2 + \frac{675}{64}e^4 + \frac{45}{128}e^6\right)$$

For example, when $-n = -7$ and $m = 1$, the look-up Table 2.2 is as follows.

n	m	Power exponent		Coefficients					Remark
		$(1-e^2)$	e	e^0	e^2	e^4	e^6	e^8	
7	1	−11/2	1	5/2	15/4	5/16			

Therefore,

$$X_0^{-7,1} = \left(1-e^2\right)^{-11/2} e\left(\frac{5}{2} + \frac{15}{4}e^2 + \frac{5}{16}e^4\right)$$

Table 2.2 Series expressions of $X_0^{-n,m}$ $(2 \le n \le 10, 0 \le m \le n-2)$

n	m	Power exponent		Coefficients					Remark
		$(1-e^2)$	e	e^0	e^2	e^4	e^6	e^8	
2	0	−1/2	0	1					
3	0	−3/2	0	1					G_{210}
3	1	−3/2	1	1/2					
4	0	−5/2	0	1	1/2				
4	1	−5/2	1	1					G_{31-1}, G_{321}
4	2	−5/2	2	1/4					
5	0	−7/2	0	1	3/2				G_{420}
5	1	−7/2	1	3/2	3/8				
5	2	−7/2	2	3/4					G_{41-2}, G_{432}
5	3	−7/2	3	1/8					
6	0	−9/2	0	1	3	3/8			
6	1	−9/2	1	2	3/2				G_{52-1}, G_{531}
6	2	−9/2	2	3/2	1/4				
6	3	−9/2	3	1/2					G_{51-3}, G_{543}
6	4	−9/2	4	1/16					
7	0	−11/2	0	1	5	15/8			G_{630}
7	1	−11/2	1	5/2	15/4	5/16			
7	2	−11/2	2	5/2	5/4				G_{62-2}, G_{642}
7	3	−11/2	3	5/4	5/32				
7	4	−11/2	4	5/16					G_{61-4}, G_{654}
7	5	−11/2	5	1/32					
8	0	−13/2	0	1	15/2	45/8	5/16		
8	1	−13/2	1	3	15/2	15/8			G_{73-1}, G_{741}
8	2	−13/2	2	15/4	15/4	15/64			
8	3	−13/2	3	5/2	15/16				G_{72-3}, G_{753}
8	4	−13/2	4	15/16	3/32				
8	5	−13/2	5	3/16					G_{71-5}, G_{765}
8	6	−13/2	6	1/64					
9	0	−15/2	0	1	21/2	105/8	35/16		G_{840}
9	1	−15/2	1	7/2	105/8	105/16	35/128		
9	2	−15/2	2	21/4	35/4	105/64			G_{83-2}, G_{852}
9	3	−15/2	3	35/8	105/32	21/128			
9	4	−15/2	4	35/16	21/32				G_{82-4} G_{864}
9	5	−15/2	5	21/32	7/128				
9	6	−15/2	6	7/64					G_{81-6}, G_{876}
9	7	−15/2	7	1/128					

(continued)

Table 2.2 (continued)

n	m	Power exponent		Coefficients					Remark
		$(1-e^2)$	e	e^0	e^2	e^4	e^6	e^8	
10	0	−17/2	0	1	14	105/4	35/4	35/128	
10	1	−17/2	1	4	21	35/2	35/16		G_{94-1}, G_{951}
10	2	−17/2	2	7	35/2	105/16	7/32		
10	3	−17/2	3	7	35/4	21/16			G_{93-3}, G_{963}
10	4	−17/2	4	35/8	21/8	7/64			
10	5	−17/2	5	7/4	7/16				G_{92-5}, G_{975}
10	6	−17/2	6	7/16	1/32				
10	7	−17/2	7	1/16					G_{91-7}, G_{987}
10	8	−17/2	8	1/256					

2.2 Hypergeometric Series Expressions of $X_0^{n,m}$ and $X_0^{-(n+1),m}$

2.2.1 *Hypergeometric Series Expressions of $X_0^{n,m}$ and $X_0^{-(n+1),m}$ in β^2*

McClain first obtained the hypergeometric series expressions of $X_0^{n,m}$ and $X_0^{-(n+1),m}$ in β^2 [4]:

$$X_0^{-(n+1),m} = \left(1-\beta^2\right)^{-(n-1)} \cos^{-n}\phi \binom{n-1}{m} \beta^m F\left(m-n+1, 1-n, m+1; \beta^2\right) \tag{2.10}$$

$$X_0^{n,m} = \left(1+\beta^2\right)^{-(n+1)} \binom{n+m+1}{m} (-\beta)^m F\left(m-n-1, -n-1, m+1; \beta^2\right) \tag{2.11}$$

where $\beta = \frac{e}{1+\sqrt{1-e^2}}$ and $\cos\phi = \sqrt{1-e^2}$.

2.2.2 *Hypergeometric Series Expressions of $X_0^{n,m}$ and $X_0^{-(n+1),m}$ in e^2*

Using the quadratic transformation of the hypergeometric series

$$F\left(a,b,a-b+1;\beta^2\right)=\left(1+\beta^2\right)^{-a}F\left(\frac{a}{2},\frac{a+1}{2},a-b+1;\frac{4\beta^2}{\left(1+\beta^2\right)^2}\right) \tag{2.12}$$

and the relations

$$\frac{4\beta^2}{\left(1+\beta^2\right)^2}=e^2,\quad \frac{\beta}{1+\beta^2}=\frac{e}{2},\quad \frac{1-\beta^2}{1+\beta^2}=\cos\phi=\sqrt{1-e^2} \tag{2.13}$$

the hypergeometric series expressions of $X_0^{n,m}$ and $X_0^{-(n+1),m}$ in e^2 are derived as [4]

$$X_0^{-(n+1),m}=\cos^{-(2n-1)}\phi\left(\frac{e}{2}\right)^m\binom{n-1}{m}F\left(\frac{m-n+1}{2},\frac{m-n+2}{2},m+1;e^2\right) \tag{2.14}$$

$$X_0^{n,m}=\left(-\frac{e}{2}\right)^m\binom{n+m+1}{m}F\left(\frac{m-n-1}{2},\frac{m-n}{2},m+1;e^2\right) \tag{2.15}$$

2.2.3 Comparison with Series Expressions

A hypergeometric series is defined as

$$F(a,b,c;z)=\sum_{l=0}^{\infty}\frac{(a)_l(b)_l}{(c)_l l!}z^l \tag{2.16}$$

where $(a)_l$ is the Pochhammer symbol and $(a)_l=a(a+1)(a+2)\cdots(a+l-1)$. Because the numbers a and b of the hypergeometric series in the expressions in Eqs. (2.10) and (2.11) for $X_0^{-(n+1),m}$ and $X_0^{n,m}$ are negative, the corresponding hypergeometric series are finite ones.

(1) For $X_0^{n,m}$:

We first determine the upper limit of the summation of the hypergeometric series in the expression of $X_0^{n,m}$. According to Eq. (2.15), in this case, we have

$$a=\frac{m-n-1}{2},\quad b=\frac{m-n}{2}$$

If $n-m=2k$, then $(b)_{k+1}=(-k)_{k+1}=(-k)_k(-k+k)=0$ and the upper limit of the summation is $k=\frac{n-m}{2}$; if $n-m+1=2k$, then $(a)_{k+1}=(-k)_{k+1}=(-k)_k(-k+k)=0$ and the upper limit of the summation is $k=\frac{n-m+1}{2}$. Therefore, the upper limit of the summation of the hypergeometric series is unified as $k=$

$\left[\frac{n-m+1}{2}\right]$, and

$$F\left(\frac{m-n-1}{2}, \frac{m-n}{2}, m+1; e^2\right) = \sum_{l=0}^{[(n+1-m)/2]} \frac{\left(\frac{m-n-1}{2}\right)_l \left(\frac{m-n}{2}\right)_l}{(m+1)_l l!} e^{2l}$$

$$= \sum_{l=0}^{[(n+1-m)/2]} \frac{(m-n-1)_{2l}}{(m+1)_l l!} \left(\frac{e}{2}\right)^{2l}$$

$$= \sum_{l=0}^{[(n+1-m)/2]} \frac{(m-n-1)(m-n)\cdots(m-n-1+2l-1)}{(m+1)_l l!} \left(\frac{e}{2}\right)^{2l}$$

$$= \sum_{l=0}^{[(n+1-m)/2]} \frac{(n-m+1)(n-m)\cdots(n-m-2l+2)}{(m+1)_l l!} \left(\frac{e}{2}\right)^{2l}$$

$$= \sum_{l=0}^{[(n+1-m)/2]} \frac{(n-m+1)!}{(m+1)_l l!(n-m+1-2l)!} \left(\frac{e}{2}\right)^{2l} \tag{2.17}$$

Then, by substituting Eq. (2.17) into Eq. (2.15), we obtain

$$\begin{aligned} X_0^{n,m} &= \left(-\frac{e}{2}\right)^m \binom{n+m+1}{m} F\left(\frac{m-n-1}{2}, \frac{m-n}{2}, m+1; e^2\right) \\ &= \left(-\frac{e}{2}\right)^m \binom{n+m+1}{m} \sum_{l=0}^{[(n+1-m)/2]} \frac{(n-m+1)!}{(m+1)_l l!(n-m+1-2l)!} \left(\frac{e}{2}\right)^{2l} \\ &= \left(-\frac{e}{2}\right)^m \frac{(n+m+1)!}{(n+1)!} \sum_{l=0}^{[(n+1-m)/2]} \frac{(n-m+1)!}{(m+l)!l!(n-m+1-2l)!} \left(\frac{e}{2}\right)^{2l} \\ &= (-1)^m \frac{(n+1+m)!}{(n+1)!} \sum_{l=0}^{[(n+1-m)/2]} \frac{(n+1-m)!}{l!(m+l)!(n+1-m-2l)!} \left(\frac{e}{2}\right)^{m+2l} \end{aligned} \tag{2.18}$$

This is the same as Eq. (2.5).

(B) For $X_0^{-(n+1),m}$:

Similarly,

$$F\left(\frac{m-n+1}{2}, \frac{m-n+2}{2}, m+1; e^2\right) = \sum_{l=0}^{[(n-1-m)/2]} \frac{\left(\frac{m-n+1}{2}\right)_l \left(\frac{m-n+2}{2}\right)_l}{(m+1)_l l!} e^{2l}$$

$$= \sum_{l=0}^{[(n-1-m)/2]} \frac{(m-n+1)_{2l}}{(m+1)_l l!} \left(\frac{e}{2}\right)^{2l}$$

$$
\begin{aligned}
&= \sum_{l=0}^{[(n-1-m)/2]} \frac{(m-n+1)(m-n+2)\cdots(m-n+2l)}{(m+1)_l l!}\left(\frac{e}{2}\right)^{2l} \\
&= \sum_{l=0}^{[(n-1-m)/2]} \frac{(n-m-1)(n-m-2)\cdots(n-m-2l)}{(m+1)_l l!}\left(\frac{e}{2}\right)^{2l} \\
&= \sum_{l=0}^{[(n-1-m)/2]} \frac{(n-m-1)!}{(m+1)_l l!(n-m-1-2l)!}\left(\frac{e}{2}\right)^{2l}
\end{aligned}
\tag{2.19}
$$

Substituting Eq. (2.19) into Eq. (2.14) and using Eq. (2.13), we obtain

$$
\begin{aligned}
X_0^{-(n+1),m} &= \left(1-e^2\right)^{-(n-1/2)}\left(\frac{e}{2}\right)^m \binom{n-1}{m} F\left(\frac{m-n+1}{2}, \frac{m-n+2}{2}, m+1; e^2\right) \\
&= \left(1-e^2\right)^{-(n-1/2)}\left(\frac{e}{2}\right)^m \sum_{l=0}^{[(n-1-m)/2]} \frac{(n-1)!}{(m+l)!l!(n-m-1-2l)!}\left(\frac{e}{2}\right)^{2l}
\end{aligned}
\tag{2.20}
$$

Equation (2.8) can be rewritten as

$$
\begin{aligned}
X_0^{-(n+1),m} &= \left(1-e^2\right)^{-(n-1/2)}\left(\frac{e}{2}\right)^m \sum_{l=0}^{[(n-1-m)/2]} \binom{n-1}{m+2l}\binom{m+2l}{l}\left(\frac{e}{2}\right)^{2l} \\
&= \left(1-e^2\right)^{-(n-1/2)}\left(\frac{e}{2}\right)^m \sum_{l=0}^{[(n-1-m)/2]} \frac{(n-1)!}{l!(m+l)!(n-1-m-2l)!}\left(\frac{e}{2}\right)^{2l}
\end{aligned}
\tag{2.21}
$$

Equations (2.8) and (2.14) are thus the same.

2.3 Associated Legendre Polynomial Expressions of $X_0^{n,m}$ and $X_0^{-(n+1),m}$

The hypergeometric series can also be expressed with an associated Legendre polynomial:

$$
F\left(a, b, a-b+1; \beta^2\right) = \Gamma(a-b+1)\beta^{b-a}\left(1-\beta^2\right)^{-b} P_{-b}^{b-a}\left(\frac{1+\beta^2}{1-\beta^2}\right) \tag{2.22}
$$

where $\Gamma(\cdot)$ is the gamma function and $P_{-b}^{b-a}(\cdot)$ is the associated Legendre polynomial.

Substituting Eq. (2.22) into Eqs. (2.10) and (2.11), McClain obtained associated Legendre polynomial expressions of $X_0^{n,m}$ and $X_0^{-(n+1),m}$ [4]:

$$X_0^{-(n+1),m} = \frac{(n-1)!}{(n+m-1)!} x^n P_{n-1}^m(x) \tag{2.23}$$

$$X_0^{n,m} = (-1)^m x^{-(n+1)} \frac{(n-m+1)!}{(n+1)!} P_{n+1}^m(x) \tag{2.24}$$

where

$$x = \frac{1+\beta^2}{1-\beta^2} \tag{2.25}$$

The following points are noted.

(1) Here, the definition of the associated Legendre polynomial is

$$P_n^m(x) = \frac{1}{2^n n!}\left(x^2-1\right)^{m/2} \frac{d^{n+m}}{dx^{n+m}}\left(x^2-1\right)^n, \quad 1 < x < \infty \tag{2.26}$$

which is different from the commonly adopted definition.

(B) Cook presented similar associated Legendre polynomial expressions of $X_0^{n,m}$ and $X_0^{-(n+1),m}$ [5] but without the term $(-1)^m$ in Eq. (2.24). As stated by McClain [4], this is due to the different definitions of the associated Legendre polynomial.

2.4 Laplace Coefficient Expression of $X_0^{n,m}$

Using the true anomaly f as the integration variable, the Hansen coefficient $X_0^{n,m}$ is defined as [6]

$$X_0^{n,m} = \frac{1}{\sqrt{1-e^2}} \frac{1}{2\pi} \int_0^{2\pi} \left(\frac{r}{a}\right)^{n+2} \exp(jmf)df \tag{2.27}$$

Using the expression given by Hansen [6, 7],

$$\frac{r}{a} = \frac{\left(1-e^2\right)\left(1+\beta^2\right)}{(1+\beta\xi)\left(1+\beta\xi^{-1}\right)} \tag{2.28}$$

in which

$$\xi = \exp(jf), \quad \beta = \frac{e}{1+\sqrt{1-e^2}} \tag{2.29}$$

Laskar obtained the Laplace coefficient expression of $\left(\frac{r}{a}\right)^{n+2}$ [6]:

$$\left(\frac{r}{a}\right)^{n+2} = \left(1-e^2\right)^{n+2}\left(1+\beta^2\right)^{n+2}\frac{1}{2}\sum_{k=-\infty}^{\infty} b_{n+2}^{(k)}(-\beta)\xi^k \tag{2.30}$$

In Eq. (2.30), $b_s^{(k)}(\alpha)$ is the classical Laplace coefficient defined by

$$(1-\alpha z)^{-s}\left(1-\alpha z^{-1}\right)^{-s} = \frac{1}{2}\sum_{k=-\infty}^{\infty} b_s^{(k)}(\alpha)z^k \tag{2.31}$$

The Laplace coefficient has the characteristics that

$$b_s^{(k)}(-\alpha) = (-1)^k b_s^{(k)}(\alpha), \quad b_s^{(-k)}(\alpha) = b_s^{(k)}(\alpha) \tag{2.32}$$

Moreover, for $k \geq 0$, the Laplace coefficient has the hypergeometric series expression

$$b_s^{(k)}(\alpha) = \frac{(s)_k}{k!}\alpha^k F\left(s, s+k, k+1; \alpha^2\right) \tag{2.33}$$

where $(s)_k = s(s+1)\cdots(s+k-1)$ for $k \geq 1$, and $(s)_0 = 1$.

Therefore, according to Eq. (2.27), the Laplace coefficient expression of $X_0^{n,m}$ is [6]

$$X_0^{n,m} = \frac{(-1)^m}{2}\left(1-e^2\right)^{n+3/2}\left(1+\beta^2\right)^{n+2} b_{n+2}^{(m)}(\beta) \tag{2.34}$$

2.5 Recursion Relations of $X_0^{n,m}$ and $X_0^{-(n+1),m}$

2.5.1 *McClain's Recursion Formulae*

Recursion relations for the associated Legendre polynomial are [4]

$$P_{n+1}^m(x) = \frac{1}{n-m+1}\left[(2n+1)xP_n^m(x) - (n+m)P_{n-1}^m(x)\right]$$
$$P_n^{m+1}(x) = \frac{-2mx}{\sqrt{x^2-1}}P_n^m(x) + (n-m+1)(n+m)P_n^{m-1}(x)$$
$$P_{n+1}^m(x) = P_{n-1}^m(x) + (2n+1)\left(x^2-1\right)^{1/2}P_n^{m-1}(x)$$

$$P_n^m(x) = \frac{1}{x}\left[P_{n-1}^m(x) + (n-m+1)\left(x^2-1\right)^{1/2} P_n^{m-1}(x)\right] \tag{2.35}$$

Moreover, it follows from Eq. (2.23) that

$$\begin{aligned}
P_{n-1}^m(x) &= X_0^{-(n+1),m} \frac{(n+m-1)!}{(n-1)!} x^{-n} \\
P_n^m(x) &= \frac{(n+m)}{nx} X_0^{-(n+2),m} \frac{(n+m-1)!}{(n-1)!} x^{-n} \\
P_{n+1}^m(x) &= \frac{(n+m+1)(n+m)}{n(n+1)x^2} X_0^{-(n+3),m} \frac{(n+m-1)!}{(n-1)!} x^{-n} \\
P_n^{m+1}(x) &= (n+m+1) X_0^{-(n+2),m+1} \frac{(n+m)!}{n!} x^{-n-1} \\
P_n^{m-1}(x) &= \frac{1}{n+m} X_0^{-(n+2),m-1} \frac{(n+m)!}{n!} x^{-n-1} \\
x^2 &= \left(1-e^2\right)^{-1}, \quad \frac{x}{\sqrt{x^2-1}} = \frac{1}{e}
\end{aligned} \tag{2.36}$$

Substituting Eq. (2.36) into Eq. (2.35), McClain obtained the recursion formulae of $X_0^{-(n+1),m}$ [4]:

$$X_0^{-(n+1),m} = \frac{(n-1)\left(1-e^2\right)^{-1}}{(n+m-1)(n-m-1)}\left[(2n-3)X_0^{-n,m} - (n-2)X_0^{-(n-1),m}\right] \tag{2.37}$$

$$X_0^{-(n+1),m+1} = \frac{1}{n+m}\left[-\frac{2m}{e} X_0^{-(n+1),m} + (n-m)X_0^{-(n+1),m-1}\right] \tag{2.38}$$

$$X_0^{-(n+1),m} = \frac{(n-1)\left(1-e^2\right)^{-1}}{(n+m-1)(n+m-2)}\left[(2n-3)eX_0^{-n,m-1} + (n-2)X_0^{-(n-1),m}\right] \tag{2.39}$$

$$X_0^{-(n+1),m} = \frac{1}{n+m-1}\left[(n-1)X_0^{-n,m} + (n-m)eX_0^{-(n+1),m-1}\right] \tag{2.40}$$

Note that the coefficient $(2n-3)$ in Eq. (2.39) is incorrectly written as $(2n+1)$ in Ref. [4]. Of course, the use of the above recursion relations requires that

$$X_0^{-(n+1),m} \equiv 0, \quad m \geq n \tag{2.41}$$

If the initial value $X_0^{-2,0} = \left(1-e^2\right)^{-1/2}$ is known, the recursion starts from $n=2$. It is seen that Eq. (2.37) has a singularity of $(n-m-1)=0$ and Eq. (2.38) has a singularity of $e=0$. Equations (2.37) and (2.38) are thus inadequate. For Eq. (2.40), there is no initial value $X_0^{-(n+1),m-1}$ when $m=0$, and the equation cannot be used in the whole process. Although there is also a singularity of $(n+m-2)=0$, it is more effective to use Eq. (2.39). At $n \geq 3$, there is no longer a singularity, and as long as the recursion starts from $X_0^{-4,0}$, there is no difficulty in conducting the recursion calculation. However, Eq. (2.39) requires more initial values; i.e., three initial values are required:

$$X_0^{-2,0} = \left(1-e^2\right)^{-1/2}, \quad X_0^{-3,0} = \left(1-e^2\right)^{-3/2}, \quad X_0^{-3,1} = \frac{e}{2}\left(1-e^2\right)^{-3/2}$$

Of course, this is a disadvantage, and we thus need to devise better recursion methods.

It follows from Eq. (2.24) that

$$\begin{aligned}
P_{n+1}^m(x) &= \frac{n(n+1)x^2}{(n-m+1)(n-m)} X_0^{n,m}(-1)^m x^{n-1} \frac{(n-1)!}{(n-m-1)!} \\
P_n^m(x) &= \frac{nx}{n-m} X_0^{n-1,m}(-1)^m x^{n-1} \frac{(n-1)!}{(n-m-1)!} \\
P_{n-1}^m(x) &= X_0^{n-2,m}(-1)^m x^{n-1} \frac{(n-1)!}{(n-m-1)!} \\
P_n^{m+1}(x) &= -X_0^{n-1,m+1}(-1)^m x^n \frac{n!}{(n-m-1)!} \\
P_n^{m-1}(x) &= -\frac{1}{(n-m+1)(n-m)} X_0^{n-1,m-1}(-1)^m x^n \frac{n!}{(n-m-1)!}
\end{aligned} \tag{2.42}$$

Substituting Eq. (2.42) into Eq. (2.35), McClain further obtained the recursion formulae of $X_0^{n,m}$ [4]:

$$X_0^{n,m} = \frac{1}{n+1}\left[(2n+1)X_0^{n-1,m} - \frac{(n-m)(n+m)}{n}\left(1-e^2\right)X_0^{n-2,m}\right] \tag{2.43}$$

$$X_0^{n,m+1} = \frac{1}{n-m+1}\left[\frac{2m}{e}X_0^{n,m} + (n+m+1)X_0^{n,m-1}\right] \tag{2.44}$$

$$(n+1)X_0^{n,m} = \frac{(n-m)(n-m+1)}{n}\left(1-e^2\right)X_0^{n-2,m} - (2n+1)eX_0^{n-1,m-1} \tag{2.45}$$

$$X_0^{n,m} = \frac{n-m+1}{n+1}\left(1-e^2\right)X_0^{n-1,m} - eX_0^{n,m-1} \tag{2.46}$$

The recursion calculation still requires that

$$X_0^{n,m} \equiv 0, \quad m > n+1 \tag{2.47}$$

It is noted that $X_0^{n,n+1} \neq 0$. It follows from Eq. (2.5) that

$$X_0^{n,n+1} = (-1)^{n+1} C_{n+1}^{2n+2} \left(\frac{e}{2}\right)^{n+1} \tag{2.48}$$

The perturbation calculation requires Hansen coefficients up to $X_0^{n,n}$, whereas $X_0^{n,n+1}$ is required in the recursion calculation, which should be noted during application. An examination of the recursion formulae Eqs. (2.43) –(2.46) reveals the following.

(1) Using Eq. (2.43) to perform recursion calculation, the index m is invariant in the recursion. Because the index m of $X_0^{n-2,m}$ takes the values of $0, 1, \cdots, n-1$, for $X_0^{n,m}$ there are two functions $X_0^{n,n}$ and $X_0^{n,n+1}$ that cannot be recursively derived with Eq. (2.43), and other formulae are required in the calculation. The recursion with Eq. (2.43) is thus incomplete.
(2) There is a denominator e in Eq. (2.44), and the equation is thus a poor recursion formula.
(3) For the recursive processes of Eqs. (2.45) and (2.46), it is clear that the recursion with Eq. (2.46) needs to be calculated up to $m = n + 1$, starting from $n = 3$, whereas the recursion with Eq. (2.45) has no problem starting from $n = 2$, but one more dimension of the array may be required.

2.5.2 Giacaglia's Recursion Formulae

Using the definition of the Hansen coefficient and the identity relations

$$\begin{aligned}
&n\left(\frac{r}{a}\right)^{n-1}\frac{e}{\sqrt{1-e^2}}\sin f\exp(jmf) + jm\left(\frac{r}{a}\right)^{n}\exp(jmf)\sqrt{1-e^2}\left(\frac{a}{r}\right)^2\\
&=\sum_k X_k^{n,m}\, jk\exp(jkM)\\
&\left(\frac{r}{a}\right)^{n+1}\frac{a}{r}\exp(jmf)\\
&=\frac{1}{1-e^2}\left(\frac{r}{a}\right)^{n+1}\exp(jmf)\\
&\quad+\frac{e}{2(1-e^2)}\left(\frac{r}{a}\right)^{n+1}\{\exp[j(m+1)f]+\exp[j(m-1)f]\}\\
&=\sum_k X_k^{n,m}\exp(jkM)
\end{aligned} \tag{2.49}$$

Giacaglia presented the recursion formulae of $X_k^{n,m}$ [8]:

$$\begin{aligned}
kX_k^{n,m} &= \frac{ne}{2\sqrt{1-e^2}}\left(X_k^{n-1,m-1} - X_k^{n-1,m+1}\right) + m\sqrt{1-e^2}X_k^{n-2,m}\\
(1-e^2)X_k^{n,m} &= X_k^{n+1,m} + \frac{1}{2}e\left(X_k^{n+1,m+1} + X_k^{n+1,m-1}\right)
\end{aligned} \tag{2.50}$$

Equation (2.50) holds for all n, m and k. In particular, when $k = 0$, we have

$$X_0^{n,m} = \frac{(n+2)e}{2m\left(1-e^2\right)}\left(X_0^{n+1,m+1} - X_0^{n+1,m-1}\right) \tag{2.51}$$

$$\left(1-e^2\right)X_0^{n,m} = X_0^{n+1,m} + \frac{1}{2}e\left(X_0^{n+1,m+1} + X_0^{n+1,m-1}\right) \tag{2.52}$$

For $n > 0$, the above two formulae can only be used to perform a recursion calculation from high to low. Moreover, Eq. (2.51) has a singularity of $m = 0$. If the index n in Eqs. (2.51) and (2.52) is replaced by $-(n+1)$, i.e., $n \Rightarrow -(n+1)$, the recursion formulae of $X_0^{-(n+1),m}$ is obtained as

$$X_0^{-(n+1),m} = \frac{(-n+1)e}{2m\left(1-e^2\right)}\left(X_0^{-n,m+1} - X_0^{-n,m-1}\right) \tag{2.53}$$

$$\left(1-e^2\right)X_0^{-(n+1),m} = X_0^{-n,m} + \frac{1}{2}e\left(X_0^{-n,m+1} + X_0^{-n,m-1}\right) \tag{2.54}$$

There is still a singularity of $m = 0$ in Eq. (2.53). Using the initial value

$$X_0^{-2,0} = \left(1-e^2\right)^{-1/2} \tag{2.55}$$

a recursion with Eq. (2.54) can be performed. However, it is noted that $X_0^{-n,m+1} = 0$ when $m \geq n-1$ in the recursion calculation, and $X_0^{-n,m-1} = X_0^{-n,|m-1|}$ when $m = 0$.

2.5.3 *Laskar's Recursion Formulae*

According to the two recursion relations of the Laplace coefficient,

$$b_{n+1}^{(m)}(\beta) = \frac{(n+m)\left(1+\beta^2\right)}{n\left(1-\beta^2\right)^2} b_n^{(m)}(\beta) - \frac{2(m-n+1)\beta}{n\left(1-\beta^2\right)^2} b_n^{(m+1)}(\beta) \tag{2.56}$$

$$b_{n+1}^{(m+1)}(\beta) = \frac{m}{m-n}\left(\beta + \frac{1}{\beta}\right) b_{n+1}^{(m)}(\beta) - \frac{m+n}{m-n} b_{n+1}^{(m-1)}(\beta) \tag{2.57}$$

and Eq. (2.34), Eqs. (2.56) and (2.57) can be transformed to the recursion formulae of $X_0^{n,m}$ [6]:

$$X_0^{n,m} = \frac{n+m+1}{n+1} X_0^{n-1,m} + \frac{m-n}{n+1} e X_0^{n-1,m+1} \tag{2.58}$$

$$X_0^{n,m+1} = -\frac{2}{e}\frac{m}{m-n-1} X_0^{n,m} - \frac{m+n+1}{m-n-1} X_0^{n,m-1} \tag{2.59}$$

Equation (2.59) is the same as Eq. (2.44) is not a new recursion formula. In contrast, Eq. (2.58) is a new (and simple) recursion formula. However, in the calculation of $X_0^{n,n}$, although $X_0^{n-1,n+1} = 0$, we still need the function $X_0^{n-1,n}$. Moreover, $X_0^{n,n+1}$ cannot be recursively derived with Eq. (2.58), and Eq. (2.48) is needed in the calculation.

2.5.4 Recursion Formulae Based on the Hypergeometric Series

In reviewing the above recursion formulae, we realize that if there are the recursion relations

$$\begin{aligned} A_1 X_0^{n+1,m} + A_2 X_0^{n,m} + A_3 X_0^{n,m-1} &= 0 \\ B_1 X_0^{-(n+1),m} + B_2 X_0^{-n,m} + B_3 X_0^{-n,m-1} &= 0 \end{aligned}$$

then the recursive process are simple, starting from the same initial values. However, there are no such recursion formulae for the known recursion relations.

Such recursion formulae can be developed using contiguous relations of the hypergeometric series. We develop these recursion relations as follows.

(1) Recursion formula of $X_0^{n,m}$

The hypergeometric series expression of $X_0^{n,m}$ in β^2 is

$$X_0^{n,m} = \left(1+\beta^2\right)^{-(n+1)} \binom{n+m+1}{m} (-\beta)^m F\left(m-n-1, -n-1, m+1; \beta^2\right)$$

Then,

$$\begin{aligned} X_0^{n+1,m} &= A_{n+1,m} F\left(m-n-2, -n-2, m+1; \beta^2\right) \\ X_0^{n,m} &= A_{n,m} F\left(m-n-1, -n-1, m+1; \beta^2\right) \\ X_0^{n,m-1} &= A_{n,m-1} F\left(m-n-2, -n-1, m; \beta^2\right) \end{aligned} \tag{2.60}$$

where

$$\begin{aligned} A_{n+1,m} &= \frac{n+m+2}{\left(1+\beta^2\right)(n+2)} \left(1+\beta^2\right)^{-(n+1)} \frac{(n+m+1)!}{(n+1)!m!} (-\beta)^m \\ A_{n,m} &= \left(1+\beta^2\right)^{-(n+1)} \frac{(n+m+1)!}{(n+1)!m!} (-\beta)^m \\ A_{n,m-1} &= -\frac{m}{(n+m+1)\beta} \left(1+\beta^2\right)^{-(n+1)} \frac{(n+m+1)!}{(n+1)!m!} (-\beta)^m \end{aligned} \tag{2.61}$$

It is clear that the required contiguous function relation is

$$A_1 F(a-1,b-1) + A_2 F + A_3 F(a-1,c-1) = 0 \tag{2.62}$$

and the corresponding conversion formulae are

$$\begin{aligned} F(a-1,b-1) &= \tfrac{(1+\beta^2)(n+2)}{n+m+2} X_0^{n+1,m} \\ F &= X_0^{n,m} \\ F(a-1,c-1) &= -\tfrac{(n+m+1)\beta}{m} X_0^{n,m-1} \end{aligned} \tag{2.63}$$

The recursion formula given as Eq. (2.62) can be deduced using Gauss's contiguous relations of a hypergeometric series.

Considering Gauss's contiguous relation of a hypergeometric series [9]

$$[b-1-(c-a-1)z]F + (c-b)F(b-1) - (c-1)(1-z)F(c-1) = 0$$

and replacing the index a by $a-1$ (i.e., $a \Rightarrow a-1$), we have

$$\begin{aligned} &[b-1-(c-a)z]F(a-1) + (c-b) \\ &\quad \times F(a-1,b-1) - (c-1)(1-z)F(a-1,c-1) = 0 \end{aligned} \tag{2.64}$$

Similarly, considering Gauss's contiguous relation of a hypergeometric series [9]

$$(c-a-1)F + aF(a+1) - (c-1)F(c-1) = 0$$

and replacing the index a by $a-1$ (i.e., $a \Rightarrow a-1$), we have

$$(c-a)F(a-1) + (a-1)F - (c-1)F(a-1,c-1) = 0 \tag{2.65}$$

Substituting Eq. (2.65) into Eq. (2.64) to eliminate $F(a-1)$ yields

$$\begin{aligned} &(c-a)(c-b)F(a-1,b-1) \\ &-(a-1)[b-1-(c-a)z]F + (a+b-c-1)(c-1)F(a-1,c-1) = 0 \end{aligned} \tag{2.66}$$

Furthermore, substituting $a = m-n-1$, $b = -n-1$, $c = m+1$ and $z = \beta^2$ into Eq. (2.66), we obtain

$$(n+2)X_0^{n+1,m} + (m-n-2)X_0^{n,m} + (n+m+1)\frac{2\beta}{1+\beta^2}X_0^{n,m-1} = 0 \tag{2.67}$$

As $2\beta/(1+\beta^2) = e$, we ultimately obtain from Eq. (2.67) the recursion formula of $X_0^{n,m}$:

$$X_0^{n+1,m} = \frac{1}{n+2}\left[(n-m+2)X_0^{n,m} - (n+m+1)eX_0^{n,m-1}\right] \tag{2.68}$$

(B) Recursion formula of $X_0^{-(n+1),m}$

The hypergeometric series expression of $X_0^{-(n+1),m}$ in β^2 is

$$X_0^{-(n+1),m} = \left(1-\beta^2\right)^{-(n-1)} \cos^{-n}\phi \binom{n-1}{m} \beta^m F\left(m-n+1, 1-n, m+1; \beta^2\right)$$

Accordingly, we have

$$\begin{aligned} X_0^{-(n+1),m} &= A_{n+1,m} F(a-1, b-1) \\ X_0^{-n,m} &= A_{n,m} F \\ X_0^{-n,m-1} &= A_{n,m-1} F(a-1, c-1) \end{aligned} \tag{2.69}$$

where

$$\begin{aligned} A_{n+1,m} &= \frac{n-1}{\left(1-\beta^2\right)\cos\phi(n-m-1)}\left(1-\beta^2\right)^{-(n-2)} \cos^{-n+1}\phi \frac{(n-2)!}{(n-2-m)!m!}\beta^m \\ A_{n,m} &= \left(1-\beta^2\right)^{-(n-2)} \cos^{-n+1}\phi \frac{(n-2)!}{(n-2-m)!m!}\beta^m \\ A_{n,m-1} &= \frac{m}{(n-m-1)\beta}\left(1-\beta^2\right)^{-(n-2)} \cos^{-n+1}\phi \frac{(n-2)!}{(n-2-m)!m!}\beta^m \end{aligned} \tag{2.70}$$

The required contiguous relation of the hypergeometric series is again Eq. (2.62); i.e.,

$$A_1 F(a-1, b-1) + A_2 F + A_3 F(a-1, c-1) = 0$$

However, the corresponding coefficients are $a = m-n+2, b = 2-n, c = m+1$ and $z = \beta^2$. Substituting these coefficients into Eq. (2.66) and using the conversion relations

$$\begin{aligned} F(a-1, b-1) &= \tfrac{\left(1-\beta^2\right)\cos\phi(n-m-1)}{n-1} X_0^{-(n+1),m} \\ F &= X_0^{-n,m} \\ F(a-1, c-1) &= \tfrac{(n-m-1)\beta}{m} X_0^{-n,m-1} \end{aligned} \tag{2.71}$$

yields

$$(n+m-1)\frac{\left(1-\beta^2\right)\cos\phi}{(n-1)\left(1+\beta^2\right)} X_0^{-(n+1),m} - X_0^{-n,m} - \frac{2\beta}{1+\beta^2} X_0^{-n,m-1} = 0 \tag{2.72}$$

and the recursion formula of $X_0^{-(n+1),m}$ is ultimately derived as

$$\frac{n+m-1}{n-1}(1-e^2)X_0^{-(n+1),m} = X_0^{-n,m} + eX_0^{-n,m-1} \tag{2.73}$$

In this way, we obtain the two recursion formulae Eqs. (2.68) and (2.73). Adding the initial values and the ranges of the indexes of the recursions, the complete recursion formulae are obtained as follows [10].

For $X_0^{n,m}$,

$$\begin{aligned} &X_0^{n+1,m} = \tfrac{1}{n+2}\left[(n-m+2)X_0^{n,m} - (n+m+1)eX_0^{n,m-1}\right] \\ &X_0^{0,0} = 1, \quad X_0^{0,1} = -e, \quad (n = 0, 1, \cdots; m = 0, 1, \cdots, n+2) \end{aligned} \tag{2.74}$$

For $X_0^{-(n+1),m}$,

$$\begin{aligned} &\tfrac{n+m-1}{n-1}(1-e^2)X_0^{-(n+1),m} = X_0^{-n,m} + eX_0^{-n,m-1} \\ &X_0^{-2,0} = (1-e^2)^{-1/2}, \quad (n = 2, 3, \cdots; m = 0, 1, \cdots, n-1) \end{aligned} \tag{2.75}$$

2.5.5 Recursion Formulae of $X_0^{n,m}$ and $X_0^{-(n+1),m}$ Used in the Perturbation Calculations

There are two eccentricity functions in satellite dynamics. Their relations with the Hansen coefficient $X_k^{n,m}$ are [11, 12]

$$\begin{aligned} G_{lpq}(e) &= X_{l-2p+q}^{-(l+1),l-2p} \\ H_{lpq}(e) &= X_{l-2p+q}^{l,l-2p} \end{aligned}$$

The eccentricity functions related to $X_0^{n,m}$ are $G_{lp(2p-l)}$ and $H_{lp(2p-l)}$; i.e., only the Hansen coefficients with $m = l - 2p$ need to be calculated in satellite dynamics, which means that the computational cost is less than half that of the above recursion calculation of $X_0^{n,m}$. Scholars have thus studied the recursion calculation of $G_{lp(2p-l)}$ and $H_{lp(2p-l)}$, with the hope of usefulness of all the calculated eccentricity functions.

Vakhidov presented three recursion formulae for $X_0^{n,m}$ in the case that $n > 0$ [13]:

$$\begin{aligned} 2e(n+3)X_0^{n+2,m-2} = &\left[4(1-m) - e^2(9-3m+2n)\right]X_0^{n+1,m-1} \\ &+ e^2(m-5-2n)X_0^{n+1,m+1} \\ &+ 2e(n+3)(1-e^2)X_0^{n,m}, \quad (m = -n) \end{aligned} \tag{2.76}$$

$$
\begin{aligned}
2(n+3)X_0^{n+2,m} &= e(-m-5-2n)X_0^{n+1,m-1} + e(m-5-2n)X_0^{n+1,m+1} \\
&\quad +2\left(1-e^2\right)(n+3)X_0^{n,m}, \qquad (m \neq \pm n)
\end{aligned}
\tag{2.77}
$$

$$
\begin{aligned}
2e(n+3)X_0^{n+2,m+2} &= \left[4(1+m) - e^2(9+3m+2n)\right]X_0^{n+1,m+1} \\
&\quad - e^2(m+5+2n)X_0^{n+1,m-1} \\
&\quad + 2e(n+3)\left(1-e^2\right)X_0^{n,m}, \quad (m=n)
\end{aligned}
\tag{2.78}
$$

However, Vakhidov did not clearly state that Eq. (2.77) is not true for $m = \pm n$. Moreover, because Vakhidov adopted a computer algebra calculation, the condition $X_0^{n+2,-(n+2)} = X_0^{n+2,n+2}$ was not used, and Eqs. (2.76) and (2.78) with the denominator e were obtained. Therefore, this set of recursion formulae is poor.

Making a small change, Eq. (2.77) can be applied to the case that $m = \pm n$. From Eq. (2.51), we have

$$
X_0^{n+1,m+1} = \frac{2m\left(1-e^2\right)}{(n+2)e} X_0^{n,m} + X_0^{n+1,m-1} \tag{2.79}
$$

Substituting this result into Eq. (2.77), we obtain

$$
X_0^{n+2,m} = \frac{(n-m+2)(n-m+3)}{(n+2)(n+3)}\left(1-e^2\right)X_0^{n,m} - \frac{2n+5}{n+3} e X_0^{n+1,m-1} \tag{2.80}
$$

Equation (2.80) is a recursion formula that can be applied to $-n-2 \leq m \leq n+2$ (from low to high). In fact, Eqs. (2.80) and (2.45) are the same. For $X_0^{-(n+1),m}$, the recursion formula is Eq. (2.39). The recursive processes with the initial values and ranges of the indexes are as follows.

(1) For $X_0^{n,m}$, the recursion formula is

$$
\begin{aligned}
X_0^{n+2,m} &= \tfrac{(n-m+2)(n-m+3)}{(n+2)(n+3)}\left(1-e^2\right)X_0^{n,m} - \tfrac{2n+5}{n+3} e X_0^{n+1,m-1} \\
&\quad [n = 0, 1, \cdots; m = \operatorname{mod}(n,2), \operatorname{mod}(n,2)+2, \cdots, n+2]
\end{aligned}
\tag{2.81}
$$

and the initial values are

$$
X_0^{0,0} = 1, \qquad X_0^{1,1} = -\frac{3}{2}e \tag{2.82}
$$

Note that when $m > n+1$, $X_0^{n,m} \equiv 0$; and when $m = 0$, $X_0^{n+1,m-1} = X_0^{n+1,|m-1|}$ in the recursion calculation.

(B) For $X_0^{-(n+1),m}$, the recursion formula is

$$
\begin{aligned}
X_0^{-(n+1),m} &= \tfrac{(n-1)\left(1-e^2\right)^{-1}}{(n+m-1)(n+m-2)}\left[(2n-3)eX_0^{-n,m-1} + (n-2)X_0^{-(n-1),m}\right], \\
&[n = 3, 4, \cdots; m = \operatorname{mod}(n,2), \operatorname{mod}(n,2)+2, \cdots, n-2]
\end{aligned}
\tag{2.83}
$$

and the initial value is

$$X_0^{-3,0} = \left(1 - e^2\right)^{-3/2} \tag{2.84}$$

We recall that $X_0^{-(n+1),m} \equiv 0$ when $m \geq n$ and $X_0^{-n,m-1} = X_0^{-n,|m-1|}$ when $m = 0$.

The above recursive processes are much simpler than Vakhidov's methods [13] and there is no singularity. All the obtained results are required for perturbation calculations, and there is no "useless intermediate function". If the order N is given, all the required eccentricity functions can be calculated from the one call. If the results are recorded in a two-dimensional array $X(0:N, 0:N)$, the eccentricity functions are

$$\begin{aligned} G_{lp(2p-l)} &= X(l+1, |l-2p|) \\ H_{lp(2p-l)} &= X(l, |l-2p|) \end{aligned} \tag{2.85}$$

2.6 Calculation of $X_0^{n,m}$ and $X_0^{-(n+1),m}$ and their Derivatives

2.6.1 *Direct Calculation Method*

Differentiating the series expressions of $X_0^{n,m}$ and $X_0^{-(n+1),m}$ with respect to eccentricity, direct calculation formulae of $X_0^{n,m}$ and $X_0^{-(n+1),m}$ and their derivatives are obtained as

$$\begin{aligned} X_0^{n,m} &= \left(-\frac{e}{2}\right)^m \sum_{l=0}^{[(n+1-m)/2]} \binom{n+1+m}{m}\binom{2l}{l}\binom{n+1-m}{2l}\left[\binom{l+m}{m}\right]^{-1}\left(\frac{e}{2}\right)^{2l} \\ \frac{dX_0^{n,m}}{de} &= \left(-\frac{e}{2}\right)^m \sum_{l=0}^{[(n+1-m)/2]} \binom{n+1+m}{m}\binom{2l}{l}\binom{n+1-m}{2l}\left[\binom{l+m}{m}\right]^{-1} \\ &\quad \times \frac{m+2l}{2}\left(\frac{e}{2}\right)^{2l-1} \end{aligned} \tag{2.86}$$

$$\begin{aligned}
X_0^{-(n+1),m} &= \left(1-e^2\right)^{-(n-1/2)} \sum_{k=0}^{[(n-1-m)/2]} \binom{n-1}{m+2k}\binom{m+2k}{k}\left(\tfrac{e}{2}\right)^{m+2k} \\
\frac{dX_0^{-(n+1),m}}{de} &= 2e\left(n-\tfrac{1}{2}\right)\left(1-e^2\right)^{-(n+1/2)} \sum_{k=0}^{[(n-1-m)/2]} \binom{n-1}{m+2k}\binom{m+2k}{k}\left(\tfrac{e}{2}\right)^{m+2k} \\
&\quad +\left(1-e^2\right)^{-(n-1/2)} \sum_{k=0}^{[(n-1-m)/2]} \binom{n-1}{m+2k}\binom{m+2k}{k}\tfrac{m+2k}{2}\left(\tfrac{e}{2}\right)^{m+2k-1}
\end{aligned} \tag{2.87}$$

Because Eqs. (2.86) and (2.87) are both finite series and not alternating series, there is no problem in using the direct method to calculate $X_0^{n,m}$ and $X_0^{-(n+1),m}$, but the calculation time is long.

2.6.2 Recursion Calculation Methods

In perturbation calculations, we also need to calculate the derivatives of $X_0^{n,m}$ and $X_0^{-(n+1),m}$, which can be done at the same time as the calculation of $X_0^{n,m}$ and $X_0^{-(n+1),m}$. We introduce four recursion calculation methods below. There are many "intermediate functions" when using the first two methods to calculate all $X_0^{n,m}$ and $X_0^{-(n+1),m}$. Only the useful $X_0^{n,m}$ and $X_0^{-(n+1),m}$ for perturbation calculations are calculated without any "intermediate function" when using the last two methods, but the formulae are more complex. The computational efficiencies of these methods are further compared in Sect. 2.8.

(1) Recursion calculation of $X_0^{n,m}$ and its derivative over the whole course

Differentiating Eq. (2.74) with respect to eccentricity, the recursion formulae for calculating $X_0^{n,m}$ and its derivative simultaneously are obtained as

$$\begin{aligned}
X_0^{n+1,m} &= \tfrac{1}{n+2}\left[(n-m+2)X_0^{n,m} - (n+m+1)eX_0^{n,m-1}\right] \\
\frac{dX_0^{n+1,m}}{de} &= \tfrac{1}{n+2}\left[(n-m+2)\frac{dX_0^{n,m}}{de} - (n+m+1)e\frac{dX_0^{n,m-1}}{de} - (n+m+1)X_0^{n,m-1}\right] \\
X_0^{0,0} &= 1, \quad X_0^{0,1} = -e \\
\frac{dX_0^{0,0}}{de} &= 0, \quad \frac{dX_0^{0,1}}{de} = -1, \qquad (n=0,1,\cdots,N-1;\ m=0,1,\cdots,n+2)
\end{aligned} \tag{2.88}$$

(B) Recursion calculation of $X_0^{-(n+1),m}$ and its derivative over the whole course

Differentiating Eq. (2.75) with respect to eccentricity, the recursion formulae for calculating $X_0^{-(n+1),m}$ and its derivative simultaneously are obtained as

$$\begin{aligned}
&\tfrac{n+m-1}{n-1}\left(1-e^2\right)X_0^{-(n+1),m} = X_0^{-n,m} + eX_0^{-n,m-1}\\
&\tfrac{n+m-1}{n-1}\left(1-e^2\right)\frac{dX_0^{-(n+1),m}}{de} = \frac{dX_0^{-n,m}}{de} + e\frac{dX_0^{-n,m-1}}{de} + 2e\tfrac{n+m-1}{n-1}X_0^{-(n+1),m} + X_0^{-n,m-1}\\
&X_0^{-2,0} = \left(1-e^2\right)^{-1/2},\quad \frac{dX_0^{-2,0}}{de} = e\left(1-e^2\right)^{-3/2},\quad (n=2,3,\cdots,N;\ m=0,1,\cdots,n-1)
\end{aligned} \tag{2.89}$$

(C) Recursion calculation of useful $X_0^{n,m}$ and its derivative in the perturbations

Differentiating Eq. (2.81) with respect to eccentricity, the recursion formulae for calculating useful $X_0^{n,m}$ and its derivative in the perturbations simultaneously are obtained as

$$\begin{aligned}
X_0^{n+2,m} &= \tfrac{(n-m+2)(n-m+3)}{(n+2)(n+3)}\left(1-e^2\right)X_0^{n,m} - \tfrac{2n+5}{n+3}eX_0^{n+1,m-1}\\
\frac{dX_0^{n+2,m}}{de} &= \tfrac{(n-m+2)(n-m+3)}{(n+2)(n+3)}\left(1-e^2\right)\frac{dX_0^{n,m}}{de} - \tfrac{2n+5}{n+3}e\frac{dX_0^{n+1,m-1}}{de}\\
&\quad -2e\tfrac{(n-m+2)(n-m+3)}{(n+2)(n+3)}X_0^{n,m} - \tfrac{2n+5}{n+3}X_0^{n+1,m-1}\\
X_0^{0,0} &= 1,\quad X_0^{1,1} = -\tfrac{3}{2}e,\quad \frac{dX_0^{0,0}}{de} = 0,\quad \frac{dX_0^{1,1}}{de} = -\tfrac{3}{2}\\
&[n=0,1,\cdots,N-2;\ m=\operatorname{mod}(n,2),\operatorname{mod}(n,2)+2,\cdots,n+2]
\end{aligned} \tag{2.90}$$

(D) Recursion calculation of useful $X_0^{-(n+1),m}$ and its derivative in the perturbations

Differentiating Eq. (2.83) with respect to eccentricity, the recursion formulae for calculating useful $X_0^{-(n+1),m}$ and its derivative in the perturbations simultaneously are obtained as

$$\begin{aligned}
X_0^{-(n+1),m} &= \tfrac{(n-1)\left(1-e^2\right)^{-1}}{(n+m-1)(n+m-2)}\left[(2n-3)eX_0^{-n,m-1} + (n-2)X_0^{-(n-1),m}\right]\\
\left(1-e^2\right)\frac{dX_0^{-(n+1),m}}{de} &= 2eX_0^{-(n+1),m} + \tfrac{n-1}{(n+m-1)(n+m-2)}\\
&\quad\times\left[(2n-3)X_0^{-n,m-1} + (2n-3)e\frac{dX_0^{-n,m-1}}{de} + (n-2)\frac{dX_0^{-(n-1),m}}{de}\right]\\
X_0^{-3,0} &= \left(1-e^2\right)^{-3/2},\quad \frac{dX_0^{-3,0}}{de} = 3e\left(1-e^2\right)^{-5/2}\\
&[n=3,4,\cdots,N;\ m=\operatorname{mod}(n,2),\operatorname{mod}(n,2)+2,\cdots,n-2]
\end{aligned} \tag{2.91}$$

Of course, in the above calculations, we first calculate $X_0^{n,m}$ and $X_0^{-(n+1),m}$ and then calculate their derivatives.

2.7 Calculation of $X_0^{n,m}$ and $X_0^{-(n+1),m}$ in Perturbation Theory without Singularities

In satellite dynamics, to eliminate singularities in the perturbation equations, it is necessary to adopt the non-singular elements

$$\begin{aligned} &a, \quad h = e\sin(\omega+\Omega), \quad k = e\cos(\omega+\Omega) \\ &\xi = \tan(i/2)\sin\Omega, \quad \eta = \tan(i/2)\cos\Omega, \quad \lambda = M+\omega+\Omega \end{aligned} \tag{2.92}$$

In this case, the functions involved in the perturbation expressions are not $X_0^{n,m}$ and $X_0^{-(n+1),m}$ themselves but rather $Y_0^{n,m}$ and $Y_0^{-(n+1),m}$, which are defined as

$$\begin{aligned} X_0^{n,m} &= e^{|m|}Y_0^{n,m} \\ X_0^{-(n+1),m} &= e^{|m|}Y_0^{-(n+1),m} \end{aligned} \tag{2.93}$$

As the eccentricity in $Y_0^{n,m}$ and $Y_0^{-(n+1),m}$ appears in the form of e^2, we first calculate their derivatives with respect to e^2 and then obtain the derivatives of $Y_0^{n,m}$ and $Y_0^{-(n+1),m}$ with respect to e according to

$$\begin{aligned} \frac{dY_0^{n,m}}{de} &= 2e\frac{dY_0^{n,m}}{de^2} \\ \frac{dY_0^{-(n+1),m}}{de} &= 2e\frac{dY_0^{-(n+1),m}}{de^2} \end{aligned} \tag{2.94}$$

2.7.1 Direct Calculation Method

Using Eqs. (2.86), (2.87) and (2.93), the direct calculation formulae of $Y_0^{n,m}$, $Y_0^{-(n+1),m}$ and their derivatives are obtained as

$$\begin{aligned} Y_0^{n,m} &= \left(-\frac{1}{2}\right)^m \sum_{l=0}^{[(n+1-m)/2]} \binom{n+1+m}{m}\binom{2l}{l}\binom{n+1-m}{2l}\left[\binom{l+m}{m}\right]^{-1}\left(\frac{1}{2}\right)^{2l} e^{2l} \\ \frac{dY_0^{n,m}}{de^2} &= \left(-\frac{1}{2}\right)^m \sum_{l=0}^{[(n+1-m)/2]} \binom{n+1+m}{m}\binom{2l}{l}\binom{n+1-m}{2l}\left[\binom{l+m}{m}\right]^{-1}\left(\frac{1}{2}\right)^{2l} \\ &\quad \times l e^{2(l-1)} \end{aligned} \tag{2.95}$$

$$\begin{aligned} Y_0^{-(n+1),m} &= \left(1-e^2\right)^{-(n-1/2)}\frac{1}{2^m}\sum_{k=0}^{[(n-1-m)/2]}\binom{n-1}{m+2k}\binom{m+2k}{k}\left(\frac{e}{2}\right)^{2k} \\ \frac{dY_0^{-(n+1),m}}{de^2} &= \left(n-\frac{1}{2}\right)\left(1-e^2\right)^{-(n+1/2)}\frac{1}{2^m}\sum_{k=0}^{[(n-1-m)/2]}\binom{n-1}{m+2k}\binom{m+2k}{k}\left(\frac{e}{2}\right)^{2k} \\ &\quad + \frac{\left(1-e^2\right)^{-(n-1/2)}}{4}\frac{1}{2^m}\sum_{k=1}^{[(n-1-m)/2]}\binom{n-1}{m+2k}\binom{m+2k}{k}k\left(\frac{e}{2}\right)^{2(k-1)} \end{aligned} \tag{2.96}$$

Equations (2.95) and (2.96) are simpler than Eqs. (2.86) and (2.87), and there is no problem in performing the calculations.

2.7.2 *Recursion Calculation Methods*

We next introduce recursion methods for calculating $Y_0^{n,m}$ and $Y_0^{-(n+1),m}$ and their derivatives simultaneously. On the basis of Eqs. (2.74), (2.75), (2.81) and (2.83), we present the calculation methods for four cases.

(1) Recursion calculation of $Y_0^{n,m}$ and its derivative over the whole course

Substituting Eq. (2.93) into Eq. (2.74), we obtain

$$e^{|m|}Y_0^{n+1,m} = \frac{1}{n+2}\left[(n-m+2)e^{|m|}Y_0^{n,m} - (n+m+1)ee^{|m-1|}Y_0^{n,m-1}\right] \quad (2.97)$$

For $m \neq 0$, we have

$$Y_0^{n+1,m} = \frac{1}{n+2}\left[(n-m+2)Y_0^{n,m} - (n+m+1)Y_0^{n,m-1}\right]$$

For $m = 0$, we have

$$Y_0^{n+1,m} = \frac{1}{n+2}\left[(n-m+2)Y_0^{n,m} - (n+m+1)e^2Y_0^{n,m-1}\right]$$

The recursion calculation formulae are then

$$\begin{aligned}
&Y_0^{n+1,0} = Y_0^{n,0} - \tfrac{n+1}{n+2}e^2Y_0^{n,1}, \quad (m=0)\\
&\frac{dY_0^{n+1,0}}{de^2} = \frac{dY_0^{n,0}}{de^2} - \tfrac{n+1}{n+2}\left(e^2\frac{dY_0^{n,1}}{de^2} + Y_0^{n,1}\right), \quad (m=0)\\
&Y_0^{n+1,m} = \tfrac{n-m+2}{n+2}Y_0^{n,m} - \tfrac{n+m+1}{n+2}Y_0^{n,m-1}, \quad (m \neq 0)\\
&\frac{dY_0^{n+1,m}}{de^2} = \tfrac{n-m+2}{n+2}\frac{dY_0^{n,m}}{de^2} - \tfrac{n+m+1}{n+2}\frac{dY_0^{n,m-1}}{de^2}, \quad (m \neq 0)\\
&Y_0^{0,0} = 1, \quad Y_0^{0,1} = -1\\
&\frac{dY_0^{0,0}}{de^2} = 0, \quad \frac{dY_0^{0,1}}{de^2} = 0, \quad (n = 0, 1, \cdots, N-1;\ m = 0, 1, \cdots, n+2)
\end{aligned} \quad (2.98)$$

(B) Recursion calculation of $Y_0^{-(n+1),m}$ and its derivative over the whole course

Substituting Eq. (2.93) into Eq. (2.75), we obtain

$$(1-e^2)Y_0^{-(n+1),0} = Y_0^{-n,0} + e^2Y_0^{-n,1}, \quad (m=0)$$

$$(1-e^2)\frac{dY_0^{-(n+1),0}}{de^2} = \frac{dY_0^{-n,0}}{de^2} + e^2\frac{dY_0^{-n,1}}{de^2} + Y_0^{-n,1} + Y_0^{-(n+1),0}, \quad (m=0)$$

$$
\begin{aligned}
&\left(1-e^2\right)Y_0^{-(n+1),m} = \frac{n-1}{n+m-1}\left(Y_0^{-n,m}+Y_0^{-n,m-1}\right), \quad (m \neq 0)\\
&\left(1-e^2\right)\frac{\mathrm{d}Y_0^{-(n+1),m}}{\mathrm{d}e^2} = \frac{n-1}{n+m-1}\left(\frac{\mathrm{d}Y_0^{-n,m}}{\mathrm{d}e^2}+\frac{\mathrm{d}Y_0^{-n,m-1}}{\mathrm{d}e^2}\right)+Y_0^{-(n+1),m}, \quad (m \neq 0)\\
&Y_0^{-2,0} = \left(1-e^2\right)^{-1/2},\\
&\frac{\mathrm{d}Y_0^{-2,0}}{\mathrm{d}e^2} = \frac{1}{2}\left(1-e^2\right)^{-3/2}, \quad (n=2,3,\cdots,N;\, m=0,1,\cdots,n-1)
\end{aligned}
\tag{2.99}
$$

Keep in mind that $Y_0^{-n,m} \equiv 0$ and $\frac{dY_0^{-n,m}}{de^2} \equiv 0$ when $m \geq n-1$ in the recursion calculations.

(C) Recursion calculation of useful $Y_0^{n,m}$ and its derivative in the perturbations

Adopting Eq. (2.81) and a deduction similar to that above, we have

$$
\begin{aligned}
&Y_0^{n+2,0} = \left(1-e^2\right)Y_0^{n,0} - \tfrac{2n+5}{n+3}e^2Y_0^{n+1,1}, \quad (m=0)\\
&\tfrac{dY_0^{n+2,0}}{de^2} = \left(1-e^2\right)\tfrac{dY_0^{n,0}}{de^2} - Y_0^{n,0} - \tfrac{2n+5}{n+3}\left(e^2\tfrac{dY_0^{n+1,1}}{de^2}+Y_0^{n+1,1}\right), \quad (m=0)\\
&Y_0^{n+2,m} = \tfrac{(n-m+2)(n-m+3)}{(n+2)(n+3)}\left(1-e^2\right)Y_0^{n,m} - \tfrac{2n+5}{n+3}Y_0^{n+1,m-1}, \quad (m \neq 0)\\
&\tfrac{dY_0^{n+2,m}}{de^2} = \tfrac{(n-m+2)(n-m+3)}{(n+2)(n+3)}\left[\left(1-e^2\right)\tfrac{dY_0^{n,m}}{de^2} - Y_0^{n,m}\right]\\
&\qquad - \tfrac{2n+5}{n+3}\tfrac{dY_0^{n+1,m-1}}{de^2}, \quad (m \neq 0)\\
&Y_0^{0,0} = 1, \quad Y_0^{1,1} = -\tfrac{3}{2}, \quad \tfrac{dY_0^{0,0}}{de^2} = 0, \quad \tfrac{dY_0^{1,1}}{de^2} = 0\\
&[n=0,1,\cdots,N-2;\, m=\mathrm{mod}(n,2),\mathrm{mod}(n,2)+2,\cdots,n+2]
\end{aligned}
\tag{2.100}
$$

(D) Recursion calculation of useful $Y_0^{-(n+1),m}$ and its derivative in the perturbations

Substituting Eq. (2.93) into Eq. (2.83), we obtain

$$
\begin{aligned}
&\left(1-e^2\right)Y_0^{-(n+1),0} = \tfrac{1}{n-2}\left[(2n-3)e^2Y_0^{-n,1}+(n-2)Y_0^{-(n-1),0}\right], \quad (m=0)\\
&\left(1-e^2\right)\tfrac{\mathrm{d}Y_0^{-(n+1),0}}{\mathrm{d}e^2} = \tfrac{2n-3}{n-2}\left(e^2\tfrac{\mathrm{d}Y_0^{-n,1}}{\mathrm{d}e^2}+Y_0^{-n,1}\right)+\tfrac{\mathrm{d}Y_0^{-(n-1),0}}{\mathrm{d}e^2}+Y_0^{-(n+1),0}, \quad (m=0)\\
&\left(1-e^2\right)Y_0^{-(n+1),m} = \tfrac{n-1}{(n+m-1)(n+m-2)}\\
&\qquad \times\left[(2n-3)Y_0^{-n,m-1}+(n-2)Y_0^{-(n-1),m}\right], \quad (m \neq 0)\\
&\left(1-e^2\right)\tfrac{\mathrm{d}Y_0^{-(n+1),m}}{\mathrm{d}e^2} = \tfrac{n-1}{(n+m-1)(n+m-2)}\left[(2n-3)\tfrac{\mathrm{d}Y_0^{-n,m-1}}{\mathrm{d}e^2}+(n-2)\tfrac{\mathrm{d}Y_0^{-(n-1),m}}{\mathrm{d}e^2}\right]\\
&\qquad +Y_0^{-(n+1),m}, \quad (m \neq 0)\\
&Y_0^{-3,0} = \left(1-e^2\right)^{-3/2}, \quad \tfrac{\mathrm{d}Y_0^{-3,0}}{\mathrm{d}e^2} = \tfrac{3}{2}\left(1-e^2\right)^{-5/2}\\
&[n=3,4,\cdots,N;\, m=\mathrm{mod}(n,2),\mathrm{mod}(n,2)+2,\cdots,n-2]
\end{aligned}
\tag{2.101}
$$

(E) Further analysis of the recursion calculation of $Y_0^{-(n+1),m}$

Owing to the factor $\left(1-e^2\right)^{-(n-1/2)}$ in the expression for $Y_0^{-(n+1),m}$, it seems convenient to introduce the recursion function $Z_0^{-(n+1),m}$, which is defined by

$$Y_0^{-(n+1),m} = \left(1-e^2\right)^{-(n-1/2)} Z_0^{-(n+1),m} \tag{2.102}$$

For comparison, substituting Eq. (2.102) into Eq. (2.99), we obtain the recursion formulae of $Z_0^{-(n+1),m}$:

$$\begin{aligned}
&Z_0^{-(n+1),0} = Z_0^{-n,0} + e^2 Z_0^{-n,1}, \quad (m=0)\\
&\frac{dZ_0^{-(n+1),0}}{de^2} = \frac{dZ_0^{-n,0}}{de^2} + e^2\frac{dZ_0^{-n,1}}{de^2} + Z_0^{-n,1}, \quad (m=0)\\
&\frac{n+m-1}{n-1} Z_0^{-(n+1),m} = Z_0^{-n,m} + Z_0^{-n,m-1}, \quad (m \neq 0)\\
&\frac{n+m-1}{n-1}\frac{dZ_0^{-(n+1),m}}{de^2} = \frac{dZ_0^{-n,m}}{de^2} + \frac{dZ_0^{-n,m-1}}{de^2}, \quad (m \neq 0)\\
&Z_0^{-2,0} = 1, \quad \frac{dZ_0^{-2,0}}{de^2} = 0, \quad (n = 2, 3, \cdots, N;\ m = 0, 1, \cdots, n-1)
\end{aligned} \tag{2.103}$$

Similarly, substituting Eq. (2.102) into Eq. (2.101), we obtain

$$\begin{aligned}
&Z_0^{-(n+1),0} = \left(1-e^2\right) Z_0^{-(n-1),0} + \tfrac{2n-3}{n-2} e^2 Z_0^{-n,1}, \quad (m=0)\\
&\frac{dZ_0^{-(n+1),0}}{de^2} = \left(1-e^2\right)\frac{dZ_0^{-(n-1),0}}{de^2} - Z_0^{-(n-1),0} + \tfrac{2n-3}{n-2}\left(e^2\frac{dZ_0^{-n,1}}{de^2} + Z_0^{-n,1}\right), \quad (m=0)\\
&Z_0^{-(n+1),m} = \frac{n-1}{(n+m-1)(n+m-2)}\\
&\qquad \times\left[(2n-3) Z_0^{-n,m-1} + (n-2)\left(1-e^2\right) Z_0^{-(n-1),m}\right], \quad (m \neq 0)\\
&\frac{dZ_0^{-(n+1),m}}{de^2} = \frac{n-1}{(n+m-1)(n+m-2)}\left[(2n-3)\frac{dZ_0^{-n,m-1}}{de^2}\right.\\
&\qquad \left. + (n-2)\left(1-e^2\right)\frac{dZ_0^{-(n-1),m}}{de^2} - (n-2) Z_0^{-(n-1),m}\right], \quad (m \neq 0)\\
&Z_0^{-3,0} = 1, \ \frac{dZ_0^{-3,0}}{de^2} = 0\\
&\qquad [n = 3, 4, \cdots, N;\ m = \mathrm{mod}(n,2), \mathrm{mod}(n,2)+2, \cdots, n-2]
\end{aligned} \tag{2.104}$$

Keep in mind that $Z_0^{-n,m} \equiv 0$ and $\frac{dZ_0^{-n,m}}{de^2} \equiv 0$ when $m \geq n-1$ in the recursion calculations.

Of course, when the recursion calculations are completed, $Z_0^{-(n+1),m}$ and $dZ_0^{-(n+1),m}/de^2$ need to be transformed to $Y_0^{-(n+1),m}$ and $dY_0^{-(n+1),m}/de^2$ respectively, and the corresponding calculation formulae are

$$\begin{aligned}
&Y_0^{-(n+1),m} = \left(1-e^2\right)^{-(n-1/2)} Z_0^{-(n+1),m}\\
&\frac{dY_0^{-(n+1),m}}{de^2} = \left(1-e^2\right)^{-(n+1/2)}\left[\left(1-e^2\right)\frac{dZ_0^{-(n+1),m}}{de^2} + \left(n-\tfrac{1}{2}\right) Z_0^{-(n+1),m}\right]
\end{aligned} \tag{2.105}$$

2.8 Comparison of Various Methods

To illustrate the effectiveness and computational efficiencies of the above algorithms, we compiled 16 Fortran programs (see Appendix D.1) for testing. The application scopes, calculated functions and adopted formulae are listed in Table 2.3.

Tests conducted using the compiled programs reveal the following.

(1) For $e = 0.1$, when calculating up to $n = 10$, all the results of calculating the same functions have the same 15 significant digits. For $e = 0.1$, $e = 0.5$ and $e = 0.75$, there are 14 significant digits for all the results when calculating up to $n = 100$, and the maximum relative errors of $X_0^{n,m}$, $\frac{dX_0^{n,m}}{de}$, $X_0^{-(n+1),m}$ and $\frac{dX_0^{-(n+1),m}}{de}$ (or $Y_0^{n,m}$, $\frac{dY_0^{n,m}}{de^2}$, $Y_0^{-(n+1),m}$ and $\frac{dY_0^{-(n+1),m}}{de^2}$) are all less than 0.3×10^{-14}. The methods are thus stable and the significant digits can be guaranteed.

(2) The programs that calculate only useful functions in the perturbations provide correct results.

(3) To compare the computational efficiencies of various algorithms, we tested the computing times of the complied programs for functions of order 2 –100. Because the computing of the programs is so fast that the computer cannot record the computing time of a one-time calculation precisely, we tested the computing time for 10000-time calculations. The results are listed in Table 2.3. The basic configuration parameters of the computing device were an Intel®

Table 2.3 Computing times of Hansen coefficients $X_0^{n,m}$ and $X_0^{-(n+1),m}$ and their derivatives when using Fortran programs ($e = 0.1$, 2–100 orders, 10,000-time calculations)

Application scope	Calculated functions	Adopted formula	Time (s)
Kepler elements	$X_0^{n,m}$, $\frac{dX_0^{n,m}}{de}$	(2.86)	491.7031
		(2.88)	1.0469
		(2.90)	0.7656
	$X_0^{-(n+1),m}$, $\frac{dX_0^{-(n+1),m}}{de}$	(2.87)	320.6250
		(2.54)	1.3125
		(2.89)	1.3281
		(2.91)	0.7031
Non-singular elements	$Y_0^{n,m}$, $\frac{dY_0^{n,m}}{de^2}$	(2.95)	498.1094
		(2.98)	1.2344
		(2.100)	0.7500
	$Y_0^{-(n+1),m}$, $\frac{dY_0^{-(n+1),m}}{de^2}$	(2.96)	323.3750
		(7.16) in Ref. [3]	1.3438
		(2.99)	1.1406
		(2.101)	0.6719
		(2.103)	1.5156
		(2.104)	1.1719

Xeon® W-2125 CPU @ 4.00 GHz (8 CPUs), basic frequency of approximately 4.0 GHz and 32768 MB RAM.

The following conclusions are drawn from the test results.

(1) The computing times of the direct calculation methods were hundreds of times those of the recursion methods. Therefore, in practical calculations, direct calculation methods should not be adopted generally.
(2) The computing of the recursion calculation methods was fast, and the computing time of the one-time calculation for the functions of order 2–100 was less than 0.2 ms. The methods thus provide the required computing speed in various applications.
(3) To meet the requirements of perturbation calculations, it is reasonable to choose the methods that only calculate useful functions in the perturbations, and the computing speed is higher in this way. The computing speed is high for calculations made with Eqs. (2.90), (2.91), (2.100) and (2.101).
(4) The recursion algorithm of $Z_0^{-(n+1),m}$ needs to transform $Z_0^{-(n+1),m}$ to $Y_0^{-(n+1),m}$ and thus has a lower computing speed, and the calculations made with Eqs. (2.103) and (2.104) are inferior to those made with Eqs. (2.99) and (2.101). If the transformation algorithm is further improved, this recursion algorithm may be an alternative method.

In summary, there are many alternative methods of calculating Hansen coefficients with $k = 0$ and their derivatives. Various algorithms are stable and available for various eccentricities and meet the requirements of practical application.

References

1. Plummer HC (1918) An introductory treatise on dynamical astronomy. Cambridge University Press, London
2. Laskar J, Boué G (2010) Explicit expansion of the three-body disturbing function for arbitrary eccentricities and inclinations. Astron Astrophys 522:A60
3. Wu LD (2011) Orbit and detection of artificial satellites and space debris. China Science and Technology Press, Beijing (In Chinese)
4. McClain WD (1978) A recursively formulated first-order semianalytic artificial satellite theory based on the generalized method of averaging. Volume II. the explicit development of the first-order averaged equations of motion for the nonspherical gravitational and nonresonant third-body perturbations. NASA CR-156783
5. Cook GE (1973) Basic theory for PROD, a program for computing the development of satellite orbits. Celest Mech 7(3):301–314
6. Laskar J (2005) Note on the generalized Hansen and Laplace coefficients. Celest Mech Dyn Astron 91(3–4):351–356
7. Hansen PA (1855) Entwickelung der products einer potenz des radius vectors mit dem sinus oder cosinus eines vielfachen der wahren anomalie in reihen, Abhandl. d. K. S. Ges. d. Wissensch, vol IV, pp 182–281
8. Giacaglia GEO (1976) A note on Hansen's coefficients in satellite theory. Celest Mech 14(4):515–523

9. Wu LD, Wang HB, Ma JY (2012) Inclination function in satellite dynamics. Science Press, Beijing
10. Wu LD, Zhang MJ (2012) Recursive methods for computation of Hansen coefficients $X_0^{n,m}$, $X_0^{-(n+1),m}$ and their derivatives. J Astrodyn 2(4):1–10 (In Chinese)
11. Kaula WM (1961) Analysis of gravitational and geometric aspects of geodetic utilization of satellites. Geophys J Int 5(2):104–133
12. Kaula WM (1962) Development of the lunar and solar disturbing function for a close satellite. Astron J 67(5):300–303
13. Vakhidov AA (2001) Some recurrence relations between Hansen coefficients. Celest Mech Dyn Astron 81(3):177–190

Chapter 3
Direct Methods of Calculating Hansen Coefficients and their Derivatives

This chapter investigates direct methods of calculating Hansen coefficients $X_k^{n,m}$ and their derivatives for an arbitrary index k, which is an important research issue addressed in this monograph.

3.1 Direct Methods of Calculating Hansen Coefficients

3.1.1 Newcomb-Poincaré Expression

The Newcomb-Poincaré expression is the most intuitive expression of Hansen coefficients and has the form [1]

$$X_{q+s}^{n,s} = \sum_{m=0}^{\infty} \Pi_q^{2m+|q|}(n|s)e^{2m+|q|} \tag{3.1}$$

where $\Pi_q^{2m+|q|}(n|s)$ is the Newcomb operator, which is a rational polynomial in (n, s). Denoting $\Pi_q^{2m+|q|}(n|s)$ as $X_{\rho,\sigma}^{n,s}$ with the transformations

$$\rho = m + \frac{|q|+q}{2}, \quad \sigma = m + \frac{|q|-q}{2} \tag{3.2}$$

gives the relations [1]

$$\begin{aligned} \Pi_q^{2m+|q|}(n|s) &= X_{m+(|q|+q)/2,m+(|q|-q)/2}^{n,s} \\ X_{q+s}^{n,s} &= e^{|q|}\sum_{m=0}^{\infty} X_{m+(|q|+q)/2,m+(|q|-q)/2}^{n,s} e^{2m} \end{aligned} \tag{3.3}$$

L.D Wu and M.J Zhang, *Hansen Coefficients in Satellite Orbital Dynamics*, Springer Aerospace Technology, https://doi.org/10.1007/978-981-97-0456-9_3

Izsak et al. presented a recursion relation of $X^{n,s}_{\rho,\sigma}$ [1, 2]:

$$\begin{aligned}4\rho X^{n,k}_{\rho,\sigma} &= 2(2k-n)X^{n,k+1}_{\rho-1,\sigma} + (k-n)X^{n,k+2}_{\rho-2,\sigma} + (5\rho-\sigma-4+4k-n)X^{n,k}_{\rho-1,\sigma-1} \\ &\quad - 2(\rho-\sigma+k)\sum_{\tau\geq 2}(-1)^{\tau}\binom{3/2}{\tau}X^{n,k}_{\rho-\tau,\sigma-\tau} \end{aligned} \tag{3.4}$$

where the subscripts of $X^{n,s}_{\rho,\sigma}$ are non-negative integers. Using the symmetry relation $X^{n,k}_{\rho,\sigma} = X^{n,-k}_{\sigma,\rho}$, Izsak et al. obtained another recursion relation [1, 2]:

$$\begin{aligned}4\sigma X^{n,k}_{\rho,\sigma} &= -2(2k+n)X^{n,k-1}_{\rho,\sigma-1} - (k+n)X^{n,k-2}_{\rho,\sigma-2} - (\rho-5\sigma+4+4k+n)X^{n,k}_{\rho-1,\sigma-1} \\ &\quad + 2(\rho-\sigma+k)\sum_{\tau\geq 2}(-1)^{\tau}\binom{3/2}{\tau}X^{n,k}_{\rho-\tau,\sigma-\tau} \end{aligned} \tag{3.5}$$

Adding Eqs. (3.4) and (3.5) together, the terms associated with the summation with respect to τ are eliminated, giving [1]

$$\begin{aligned}4(\rho+\sigma)X^{n,k}_{\rho,\sigma} &= 2(2k-n)X^{n,k+1}_{\rho-1,\sigma} - 2(2k+n)X^{n,k-1}_{\rho,\sigma-1} \\ &\quad + (k-n)X^{n,k+2}_{\rho-2,\sigma} - (k+n)X^{n,k-2}_{\rho,\sigma-2} \\ &\quad + 2(2\rho+2\sigma-4-n)X^{n,k}_{\rho-1,\sigma-1} \end{aligned} \tag{3.6}$$

Then, through a recursion calculation with the initial value $X^{n,k}_{0,0} = 1, X^{n,k}_{1,0} = k-\frac{n}{2}$, $X^{n,k}_{0,1} = -k-\frac{n}{2}$ and all the other $X^{n,s}_{\rho,\sigma}$ can be recursively derived.

Izsak et al. further defined [1, 2]

$$J^{n,k}_{\rho,\sigma} = 2^{\rho+\sigma}\rho!\sigma!X^{n,k}_{\rho,\sigma} \tag{3.7}$$

to make $J^{n,k}_{\rho,\sigma}$ a polynomial of the indexes n and k with integer coefficients. The recursion relations of $J^{n,k}_{\rho,\sigma}$ are [1, 2]

$$\begin{aligned} &J^{n,k}_{0,0} = 1, \quad J^{n,k}_{1,0} = 2k-n \\ &J^{n,k}_{\rho,0} = (2k-n)J^{n,k+1}_{\rho-1,0} + (\rho-1)(k-n)J^{n,k+2}_{\rho-2,0} \\ &J^{n,k}_{\rho,\sigma} = -(2k+n)J^{n,k-1}_{\rho,\sigma-1} - (\sigma-1)(k+n)J^{n,k-2}_{\rho,\sigma-2} \\ &\qquad - \rho(\rho-5\sigma+4+4k+n)J^{n,k}_{\rho-1,\sigma-1} + \rho(\rho-\sigma+k)\sum_{\tau\geq 2}C_{\rho\sigma\tau}J^{n,k}_{\rho-\tau,\sigma-\tau} \end{aligned} \tag{3.8}$$

where

$$C_{\rho\sigma\tau} = (\rho - 1)(\rho - 2)\cdots(\rho - \tau + 1)(\sigma - 1)(\sigma - 2)\cdots(\sigma - \tau + 1)C_\tau$$

$$C_\tau = (-1)^\tau \binom{3/2}{\tau} 2^{2\tau-1} = 3, 2, 3, 6, 14, 36, 99, \cdots \tag{3.9}$$

In the recursive process, when the indexes ρ and σ are negative, the values of $J^{n,k}_{\rho,\sigma}$ are zero. The expressions of $J^{n,k}_{\rho,\sigma}$ with $\rho + \sigma \leq 6$ are given in Table 3.1 [2].

Given the indexes n and k, according to Table 3.1, the values of $J^{n,k}_{\rho,\sigma}$ can be calculated. It is noted that the recursion results of the recursion formula given as Eq. (3.8) are the expressions, such as those in Table 3.1, rather than numerical results. Of course, we can extend Table 3.1 using the recursion formula Eq. (3.8). Izsak et al. [2] expanded Hansen coefficients up to the eighth order in eccentricity, e^8, and Cherniack [3] further expanded them up to the twelfth order in eccentricity, e^{12}.

After the calculation of $J^{n,k}_{\rho,\sigma}$, Hansen coefficients $X^{n,k}_{k+q}$ $(-6 \leq q \leq 6)$ are calculated using the expressions in Table 3.2, and the calculation results are truncated up to the sixth order in eccentricity, e^6. In other words, there are two steps in the calculation of Hansen coefficients using the Newcomb-Poincaré expression. The first step is to recursively derive the calculation formula of $J^{n,k}_{\rho,\sigma}$ using recursion formula Eq. (3.8). According to Eq. (3.8), the program is compiled to calculate $J^{n,k}_{\rho,\sigma}$. Hansen coefficients are then calculated from the specific Newcomb-Poincaré expressions in Table 3.2. Of course, this is not straightforward. Moreover, the procedure quantity is relatively large, but the program is simple. The computations are fast because of the integer operation.

The derivatives of Hansen coefficients are obtained by differentiating the expressions in Table 3.2 with respect to the eccentricity e. As for the non-singular elements, the kernels of Hansen coefficients and their derivatives need to be calculated, and they are derived by extracting the factor $e^{|q|}$ from the expressions of Hansen coefficients $X^{n,k}_{k+q}$. In addition, the calculation is relatively simple.

Balmino [4] presented an expression of Hansen coefficients. Making a slight transformation with Yang Hui's triangle, the expression is [5]

$$X^{n,m}_{m+s} = \left(-\frac{e}{2}\right)^s \times \sum_{t=0}^{\infty} \left(\frac{e}{2}\right)^{2t} \left\{ \sum_{j=0}^{t} \sum_{p=0}^{j} \binom{n+m+1}{j-p} X_p \sum_{q=0}^{s+j} \binom{n-m+1}{s+j-q} X_q \left[\binom{k}{t-j} - \binom{k}{t-j-1} \right] \right\} \tag{3.10}$$

where

$$k = 2t - n + s - p - q - 2, \quad X_p = \frac{(m+s)^p}{p!}, \quad X_q = \frac{(m+s)^q}{q!}(-1)^q$$

In Eq. (3.10), the index s must always be positive. If $s < 0$, the symmetry of Hansen coefficients $X^{n,m}_{m+s} = X^{n,-m}_{-m-s}$ is adopted to perform calculations; i.e., both the indexes m and s in the expression change signs.

Equation (3.10) is a series expression that can be used to calculate Hansen coefficients directly. The equation has a quadruple summation and a low computing speed.

Table 3.1 Expressions of $J^{n,k}_{\rho,\sigma}$ $(\rho+\sigma \le 6)$ [2]

$$J^{n,k}_{0,0} = 1$$

$$J^{n,k}_{1,0} = 2k - n$$

$$J^{n,k}_{2,0} = -3n + n^2 + 5k - 4kn + 4k^2$$

$$J^{n,k}_{1,1} = n + n^2 - 4k^2$$

$$J^{n,k}_{3,0} = -17n + 9n^2 - n^3 + 26k - 33kn + 6kn^2 + 30k^2 - 12k^2n + 8k^3$$

$$J^{n,k}_{2,1} = 3n + n^2 - n^3 - 2k + 5kn + 2kn^2 - 10k^2 + 4k^2n - 8k^3$$

$$\begin{aligned} J^{n,k}_{4,0} &= -142n + 95n^2 - 18n^3 + n^4 + 206k - 330kn + 102kn^2 - 8kn^3 \\ &\quad + 283k^2 - 192k^2n + 24k^2n^2 + 120k^3 - 32k^3n + 16k^4 \end{aligned}$$

$$\begin{aligned} J^{n,k}_{3,1} &= 22n - n^2 - 6n^3 + n^4 - 22k + 47kn + 3kn^2 - 4kn^3 \\ &\quad - 64k^2 + 48k^2n - 60k^3 + 16k^3n - 16k^4 \end{aligned}$$

$$J^{n,k}_{2,2} = 2n - n^2 - 2n^3 + n^4 - 9k^2 - 8k^2n^2 + 16k^4$$

$$\begin{aligned} J^{n,k}_{5,0} &= -1569n + 1220n^2 - 305n^3 + 30n^4 - n^5 + 2194k - 4080kn \\ &\quad + 1660kn^2 - 230kn^3 + 10kn^4 + 3360k^2 - 2995k^2n + 660k^2n^2 \\ &\quad - 40k^2n^3 + 1790k^3 - 840k^3n + 80k^3n^2 + 400k^4 - 80k^4n + 32k^5 \end{aligned}$$

$$\begin{aligned} J^{n,k}_{4,1} &= 231n - 68n^2 - 41n^3 + 14n^4 - n^5 - 258k + 572kn - 76kn^2 - 42kn^3 \\ &\quad + 6kn^4 - 648k^2 + 617k^2n - 60k^2n^2 - 8k^2n^3 - 614k^3 + 296k^3n \\ &\quad - 16k^3n^2 - 240k^4 + 48k^4n - 32k^5 \end{aligned}$$

$$\begin{aligned} J^{n,k}_{3,2} &= 3n - 8n^2 - 5n^3 + 6n^4 - n^5 + 10k - 10kn - 12kn^2 + 2kn^3 \\ &\quad + 2kn^4 - 12k^2 - 9k^2n - 36k^2n^2 + 8k^2n^3 + 26k^3 - 24k^3n \\ &\quad - 16k^3n^2 + 80k^4 - 16k^4n + 32k^5 \end{aligned}$$

$$\begin{aligned} J^{n,k}_{6,0} &= -21576n + 18694n^2 - 5595n^3 + 745n^4 - 45n^5 + n^6 \\ &\quad + 29352k - 60752kn + 29535kn^2 - 5530kn^3 + 435kn^4 - 12kn^5 \\ &\quad + 48538k^2 - 51615k^2n + 15345k^2n^2 - 1680k^2n^3 + 60k^2n^4 \\ &\quad + 29835k^3 - 18860k^3n + 3240k^3n^2 - 160k^3n^3 \\ &\quad + 8660k^4 - 3120k^4n + 240k^4n^2 + 1200k^5 - 192k^5n + 64k^6 \end{aligned}$$

$$\begin{aligned} J^{n,k}_{5,1} &= 3096n - 1466n^2 - 255n^3 + 185n^4 - 25n^5 + n^6 \\ &\quad - 3608k + 8454kn - 2280kn^2 - 340kn^3 + 130kn^4 - 8kn^5 \\ &\quad - 8588k^2 + 9535k^2n - 1765k^2n^2 - 80k^2n^3 + 20k^2n^4 \\ &\quad - 8200k^3 + 5280k^3n - 640k^3n^2 \\ &\quad - 3740k^4 + 1360k^4n - 80k^4n^2 - 800k^5 + 128k^5n - 64k^6 \end{aligned}$$

(continued)

Table 3.1 (continued)

$$J_{4,2}^{n,k} = -72n - 58n^2 + 5n^3 + 41n^4 - 13n^5 + n^6$$
$$+ 136k - 344kn - 115kn^2 + 66kn^3 + 17kn^4 - 4kn^5$$
$$+ 314k^2 - 399k^2n - 255k^2n^2 + 112k^2n^3 - 4k^2n^4$$
$$+ 593k^3 - 452k^3n - 136k^3n^2 + 32k^3n^3$$
$$+ 788k^4 - 304k^4n - 16k^4n^2 + 400k^5 - 64k^5n + 64k^6$$

$$J_{3,3}^{n,k} = 24n - 26n^2 - 15n^3 + 25n^4 - 9n^5 + n^6$$
$$- 172k^2 + 39k^2n - 93k^2n^2 + 48k^2n^3 - 12k^2n^4$$
$$+ 196k^4 - 48k^4n + 48k^4n^2 - 64k^6$$

Table 3.2 Expressions of Hansen coefficients $X_{k+q}^{n,k}$ $(-6 \le q \le 6)$

$q \ge 0$	$q < 0$
$X_{k+0}^{n,k} = J_{0,0}^{n,k} + \frac{e^2}{4} J_{1,1}^{n,k} + \frac{e^4}{64} J_{2,2}^{n,k} + \frac{e^6}{2304} J_{3,3}^{n,k}$	
$X_{k+1}^{n,k} = \frac{e}{2} J_{1,0}^{n,k} + \frac{e^3}{16} J_{2,1}^{n,k} + \frac{e^5}{384} J_{3,2}^{n,k}$	$X_{k-1}^{n,k} = \frac{e}{2} J_{1,0}^{n,-k} + \frac{e^3}{16} J_{2,1}^{n,-k} + \frac{e^5}{384} J_{3,2}^{n,-k}$
$X_{k+2}^{n,k} = \frac{e^2}{8} J_{2,0}^{n,k} + \frac{e^4}{96} J_{3,1}^{n,k} + \frac{e^6}{3072} J_{4,2}^{n,k}$	$X_{k-2}^{n,k} = \frac{e^2}{8} J_{2,0}^{n,-k} + \frac{e^4}{96} J_{3,1}^{n,-k} + \frac{e^6}{3072} J_{4,2}^{n,-k}$
$X_{k+3}^{n,k} = \frac{e^3}{48} J_{3,0}^{n,k} + \frac{e^5}{768} J_{4,1}^{n,k}$	$X_{k-3}^{n,k} = \frac{e^3}{48} J_{3,0}^{n,-k} + \frac{e^5}{768} J_{4,1}^{n,-k}$
$X_{k+4}^{n,k} = \frac{e^4}{384} J_{4,0}^{n,k} + \frac{e^6}{7680} J_{5,1}^{n,k}$	$X_{k-4}^{n,k} = \frac{e^4}{384} J_{4,0}^{n,-k} + \frac{e^6}{7680} J_{5,1}^{n,-k}$
$X_{k+5}^{n,k} = \frac{e^5}{3840} J_{5,0}^{n,k}$	$X_{k-5}^{n,k} = \frac{e^5}{3840} J_{5,0}^{n,-k}$
$X_{k+6}^{n,k} = \frac{e^6}{46,080} J_{6,0}^{n,k}$	$X_{k-6}^{n,k} = \frac{e^6}{46,080} J_{6,0}^{n,-k}$

However, it is an explicit-function expression of the Newcomb operator and should be given more attention.

3.1.2 Hill Expressions

(1) Various Hill Expressions of Hansen Coefficients

Various Hill expressions of Hansen coefficients are presented in Refs. [1, 4, 6–13].

1. Gaposchkin [6] quoted Plummer's [7] expression:

$$X_q^{nm} = \left(1 + \beta^2\right)^{-(n+1)} \sum_p J_p(qe) X_{qp}^{nm} \tag{3.11}$$

where

$$\begin{cases} X_{qp}^{nm} = (-\beta)^{q-p-m}\binom{n+1-m}{q-p-m} \\ \qquad \times F\left(q-p-n-1, -m-n-1, q-p-m+1; \beta^2\right), \\ \qquad (q-p-m>0) \\ X_{qp}^{nm} = (-\beta)^{-q+p+m}\binom{n+1+m}{-q+p+m} \\ \qquad \times F\left(-q+p-n-1, m-n-1, -q+p+m+1; \beta^2\right), \\ \qquad (q-p-m<0) \\ X_{qp}^{nm} = F\left(m-n-1, -m-n-1, 1; \beta^2\right), \quad (q-p-m=0) \\ \beta = \dfrac{e}{1+\sqrt{1-e^2}} \end{cases} \tag{3.12}$$

In Eq. (3.12), $F(\cdot)$ refers to the series expansion of the hypergeometric series and is defined as

$$F(a, b, c; z) = \sum_{n=0}^{\infty} \frac{(a)_n (b)_n}{(c)_n n!} z^n$$

2. Branham [8] quoted Plummer's [9] expression:

$$X_j^{n,k} = \left(1+\beta^2\right)^{-(n+1)} \sum_{p=-\infty}^{\infty} J_p(je) X_{j,p}^{n,k} \tag{3.13}$$

where

$$\begin{cases} X_{j,p}^{n,k} = (-\beta)^{j-p-k}\binom{n+1-k}{j-p-k} \\ \qquad \times F\left(j-p-n-1, -k-n-1, j-p-k+1; \beta^2\right), \\ \qquad (j-p-k>0) \\ X_{j,p}^{n,k} = (-\beta)^{-j+p+k}\binom{n+1+k}{-j+p+k} \\ \qquad \times F\left(-j+p-n-1, k-n-1, -j+p+k+1; \beta^2\right), \\ \qquad (j-p-k<0) \end{cases} \tag{3.14}$$

Equation (3.14) is similar to Eq. (3.12), but the expression for $j-p-k=0$ is not given. In fact, it is not necessary to specify the expression for $j-p-k=0$, and the expression for $j-p-k>0$ is then used for the case that $j-p-k=0$.

3. Breiter et al. [10] recommended Aksenov's [11] expression:

$$X_k^{\gamma,j} = \left(1+\beta^2\right)^{-(\gamma+1)} \sum_{s=-\infty}^{\infty} E_{k-s}^{\gamma,j} J_s(ke) \tag{3.15}$$

where

$$E_p^{\gamma,j} = (-\beta)^{p-j} \binom{\gamma-j+1}{p-j} F\left(-\gamma-j-1, p-\gamma-1, p-j+1; \beta^2\right), \ (p \geq j) \tag{3.16}$$

and $E_p^{\gamma,j}$ for $p < j$ is calculated using $E_p^{\gamma,j} = E_{-p}^{\gamma,-j}$.

4. Wnuk's [12] expression is

$$X_k^{n,m} = \left(1+\beta^2\right)^{-(n+1)} \sum_{t=-\infty}^{\infty} E_{k-t}^{n,m} J_t(ke) \tag{3.17}$$

where

$$E_{k-t}^{n,m} = \begin{cases} (-\beta)^{k-t-m} \sum_{s=0}^{\infty} \binom{n-m+1}{k-t-m+s} \binom{n+m+1}{s} \beta^{2s}, \\ (k-t-m \geq 0) \\ (-\beta)^{t-k+m} \sum_{s=0}^{\infty} \binom{n+m+1}{-k+t+m+s} \binom{n-m+1}{s} \beta^{2s}, \\ (k-t-m < 0) \end{cases} \tag{3.18}$$

Substituting the series expansion of the hypergeometric series $F(\cdot)$ into Eq. (3.16), Eq. (3.18) is derived immediately.

5. Giacaglia's [13] expression (see Appendix C for the specific deduction) is

$$X_k^{n,m} = \left(1+\beta^2\right)^{-n-1} \sum_{s=0}^{s_1} \sum_{t=0}^{t_1} \binom{n-m+1}{s} \binom{n+m+1}{t} (-\beta)^{s+t} J_{k-m-s+t}(ke) \tag{3.19}$$

where

$$s_1 = \begin{cases} n-m+1, & (n-m+1 \geq 0), \\ \infty, & (n-m+1 < 0), \end{cases}$$
$$t_1 = \begin{cases} n+m+1, & (n+m+1 \geq 0) \\ \infty, & (n+m+1 < 0) \end{cases} \tag{3.20}$$

Transforming to the eccentricity function from Eq. (3.19), we have

$$G_{lpq} = X_{l-2p+q}^{-l-1,l-2p} = \left(1+\beta^2\right)^l \sum_{s=0}^{s_1} \sum_{t=0}^{t_1} \binom{-2l+2p}{s} \binom{-2p}{t} (-\beta)^{s+t} \times J_{q-s+t}[(l-2p+q)e] \tag{3.21}$$

In Eq. (3.21), the upper limits s_1 and t_1 of the summation are

$$s_1 = \begin{cases} \infty, & (l \neq p), \\ 0, & (l = p), \end{cases} \quad t_1 = \begin{cases} \infty, & (p \neq 0) \\ 0, & (p = 0) \end{cases}$$

Using the binomial coefficient relation

$$\binom{-a}{s} = \frac{(-a)(-a-1)\cdots(-a-s+1)}{s!} = (-1)^s \binom{a+s-1}{s} \tag{3.22}$$

Equation (3.21) is expressed as

$$G_{lpq} = \left(1+\beta^2\right)^l \sum_{s=0}^{s_1} \sum_{t=0}^{t_1} \binom{2l-2p+s-1}{s} \binom{2p+t-1}{t} \beta^{s+t} J_{q-s+t}(ke) \tag{3.23}$$

where $k = l - 2p + q$. When $q - s + t < 0$, the Bessel function $J_{q-s+t}(ke)$ is calculated according to

$$J_{q-s+t}(ke) = (-1)^{|q-s+t|} J_{|q-s+t|}(ke)$$

6. Balmino [4] adopted a formula equivalent to Giacaglia's [13] expression:

$$X_k^{n,m} = \left(1+\beta^2\right)^{-n-1} \sum_{p=0}^{\infty} \sum_{q=0}^{\infty} \binom{n+m+1}{p} \binom{n-m+1}{q} (-\beta)^{p+q} J_{k-m+p-q}(ke) \tag{3.24}$$

Obviously, the upper limits of the summation of this formula should be the same as those in Eq. (3.19).

7. McClain's [1] expression is

$$X_t^{\pm n,s} = \sum_r W_r^{\pm n,s} B_t^r \tag{3.25}$$

where

$$B_t^r = \frac{r}{t} J_{t-r}(te), \quad (t \neq 0)$$

$$B_0^r = \begin{cases} 1, & (r = 0) \\ -\frac{e}{2}, & (|r| = 1) \\ 0, & (|r| > 1) \end{cases} \tag{3.26}$$

$$W_r^{n,s} = \left(1+\beta^2\right)^{-n}(-1)^{|r-s|}\binom{n-s}{a}\binom{n+s}{b} \\ \times \beta^{|r-s|} F\left(a-n+s, b-n-s, |r-s|+1; \beta^2\right) \tag{3.27}$$

$$W_r^{-n,s} = \left(1+\beta^2\right)^{n}\binom{n+s+a-1}{a}\binom{n-s+b-1}{b}\beta^{|r-s|} \\ \times F\left(n+s+a, n-s+b, |r-s|+1; \beta^2\right) \tag{3.28}$$

In Eqs. (3.27) and (3.28),

$$a = \frac{|r-s|+r-s}{2}, \quad b = \frac{|r-s|-(r-s)}{2} \tag{3.29}$$

The upper and lower limits of the summation of r are

$$\begin{cases} -n \le r \le n, & (n \ge 0) \\ -\infty \le r \le \infty, & (n < 0) \\ -1 \le r \le 1, & (t = 0) \end{cases} \tag{3.30}$$

Comparing these Hill expressions of Hansen coefficients, Eqs. (3.11), (3.13), (3.15) and (3.17) are equivalent in essence. Equations (3.19) and (3.24) are also equivalent. It is thus recommended to adopt Wnuk's expression and Giacaglia's expression; i.e., Eqs. (3.17) and (3.23) are regarded as representative Hill expressions of Hansen coefficients.

(2) Calculation of the Bessel Function

In the calculation of Hansen coefficients with Hill expressions, it is noted that the accuracy is seriously degraded if using the series expression of the Bessel function $J_n(x)$,

$$J_n(x) = \sum_{i=0}^{\infty} \frac{(-1)^i}{i!(n+i)!}\left(\frac{x}{2}\right)^{n+2i} \tag{3.31}$$

when x is large. The reason is that a computer typically has only 16 significant digits, and the largest term in formula Eq. (3.31) may be > 10^{18}. The truncation error of the computer is > 100, but the absolute value of $J_n(ke)$ is not > 1, and the calculation result is thus invalid.

Wnuk recommended an algorithm for computing the Bessel function $J_n(x)$ [12, 14]:

$$\begin{cases} J_0(x) = \sum\limits_{i=0}^{\infty} \frac{(-1)^i}{i!i!} \left(\frac{x}{2}\right)^{2i}, & (x < 20) \\ J_0(x) \approx \sqrt{\frac{2}{\pi x}} a \cos\left(x - \frac{\pi}{4} + b\right), & (x \geq 20) \\ J_s(x) = p_s J_{s-1}(x), & (s > 0) \end{cases} \tag{3.32}$$

where

$$\begin{aligned} a &= 1 - \frac{1}{16x^2} + \frac{53}{512x^4}, \\ b &= -\frac{1}{8x} + \frac{25}{384x^3}, \\ p_s &= \frac{1}{2s/x - p_{s+1}} \end{aligned} \tag{3.33}$$

It is noted that there are many asymptotic expansions for $J_0(x)$. The above calculation expressions of a and b are cited from Ref. [15]. A trial computation showed that in this way the equivalent accuracy of $J_0(x)$ is approximately 8 significant digits, which basically satisfies the calculation requirement of Hansen coefficients. Refer to Ref. [14] for the recursion of p_s starting from $p_{s+1} = 0$ and $s = 2N$, where N is the maximum value of the index s of $J_s(x)$ to be calculated. When $x = 0$, there is a problem of a zero denominator in the calculation of p_s in Eq. (3.33), and a special treatment is required.

When $s < 0$ or $x < 0$, the Bessel function is calculated according to [14]

$$J_{-s}(x) = J_s(-x) = (-1)^s J_s(x) \tag{3.34}$$

In addition, besides the above algorithm, this method is also suitable for calculating the Bessel function $J_0(x)$ with a definite integral when $|x|$ is large:

$$J_0(x) = \frac{1}{\pi} \int\limits_0^{\pi} \cos(x \sin\varphi) d\varphi \tag{3.35}$$

3.1.3 Hansen Expressions

(1) Various Hansen Expressions of Hansen Coefficients

1. On the basis of Hansen's original method, McClain [1] proved that the Hansen coefficients $X_k^{n,s}$ can be expressed as (see Appendix B for the specific proof)

$$X_k^{n,s} = \frac{\left(1-\beta^2\right)^{2n+3}}{\left(1+\beta^2\right)^{n+1}}(-\beta)^{|k-s|}\sum_{i=0}^{\infty} M_{i+(|k-s|+k-s)/2}N_{i+[|k-s|-(k-s)]/2}\beta^{2i} \tag{3.36}$$

where

$$\begin{aligned} M_p &= \sum_{m=0}^{p}(-1)^m\binom{n-k+p+1}{p-m}\frac{\mu^m}{m!}, \\ N_q &= \sum_{m=0}^{q}\binom{n+k+q+1}{q-m}\frac{\mu^m}{m!}, \\ \mu &= k\sqrt{1-e^2} \end{aligned} \tag{3.37}$$

2. Using the expression of Hansen coefficients presented by McClain [1] and the generalized Laguerre polynomials, Proulx and McClain [16] further obtained the two formulae (see Appendix B for the specific proof)

$$\begin{aligned} X_t^{n,s} &= \left(1-\beta^2\right)^{2n+3}\left(1+\beta^2\right)^{-(n+1)}(-\beta)^{|t-s|}\sum_{i=0}^{\infty} L_{i+a}^{(n-t+1)}(\mu)L_{i+b}^{(n+t+1)}(-\mu)\beta^{2i} \\ \mu &= \frac{t\left(1-\beta^2\right)}{1+\beta^2} = t\sqrt{1-e^2} \end{aligned} \tag{3.38}$$

and

$$\begin{aligned} X_t^{n,s} &= \left(1+\beta^2\right)^{-(n+1)}(-\beta)^{|t-s|}\sum_{i=0}^{\infty} L_{i+a}^{(n-s-i-a+1)}(\nu)L_{i+b}^{(n+s-i-b+1)}(-\nu)\beta^{2i}, \\ \nu - t\left(1+\beta^2\right)^{-1} &= \frac{1}{2}\frac{te}{\beta}, \end{aligned} \tag{3.39}$$

where

$$\begin{aligned} \beta &= \frac{e}{1+\sqrt{1-e^2}}, \\ a &= \frac{|t-s|+(t-s)}{2}, \\ b &= \frac{|t-s|-(t-s)}{2} \end{aligned} \tag{3.40}$$

In Eqs. (3.38) and (3.39), $L_m^{(j)}(x)$ denotes the generalized Laguerre polynomials defined by

$$L_m^{(j)}(x) = \sum_{k=0}^{m}(-1)^k\binom{m+j}{m-k}\frac{x^k}{k!} \tag{3.41}$$

3. Oesterwinter [17] presented the expression

$$G_{lpq} = X_{l-2p+q}^{-l-1,l-2p}(\beta) = (-\beta)^{|q|}\frac{\left(1+\beta^2\right)^l}{\left(1-\beta^2\right)^{2l-1}}\sum_{i=0}^{\infty} M_s N_{t'} \beta^{2i} \tag{3.42}$$

The definitions of the symbols M_s and $N_{t'}$ in Eq. (3.42) and their relations with the symbols in Eq. (3.38) are

$$\begin{aligned} M_s &= (-1)^s \sum_{\nu=0}^{s} \binom{\alpha-\nu}{s-\nu}\frac{b^\nu}{\nu!} = L_{i+a}^{(n-t+1)}(\mu) \\ N_{t'} &= (-1)^{t'} \sum_{\omega=0}^{t'} \binom{\gamma-\omega}{t'-\omega}\frac{(-b)^\omega}{\omega!} = L_{i+b}^{(n+t+1)}(-\mu) \end{aligned} \tag{3.43}$$

$$\begin{aligned} &s = i + \frac{1}{2}(|q|+q) = i+a, \quad t' = i + \frac{1}{2}(|q|-q) = i+b \\ &\alpha = 2l-2p+q-1 = -(n-t+1)-1, \quad \gamma = 2p-q-1 = -(n+t+1)-1 \\ &b = (l-2p+q)\sqrt{1-e^2} = \mu \end{aligned} \tag{3.44}$$

It is straightforward to prove that Eqs. (3.36), (3.38) and (3.42) are the same in essence.

4. Kaula [18] quoted Tisserand's [19] expression for eccentricity functions:

$$G_{lpq}(e) = (-1)^{|q|}\left(1+\beta^2\right)^l \beta^{|q|} \sum_{k=0}^{\infty} P_{lpqk} Q_{lpqk} \beta^{2k} \tag{3.45}$$

where

$$\begin{aligned} P_{lpqk} &= \sum_{r=0}^{h} \binom{2p'-2l}{h-r}\frac{(-1)^r}{r!}\left[\frac{(l-2p'+q')e}{2\beta}\right]^r \\ h &= \begin{cases} k+q', & (q'>0) \\ k, & (q'<0) \end{cases} \\ Q_{lpqk} &= \sum_{r=0}^{h} \binom{-2p'}{h-r}\frac{1}{r!}\left[\frac{(l-2p'+q')e}{2\beta}\right]^r \\ h &= \begin{cases} k, & (q'>0) \\ k-q', & (q'<0) \end{cases} \end{aligned} \tag{3.46}$$

and

$$p' = \begin{cases} p, & (p \le l/2) \\ l-p, & (p > l/2) \end{cases}$$

$$q' = \begin{cases} q, & (p \le l/2) \\ -q, & (p > l/2) \end{cases} \tag{3.47}$$

5. Breiter et al. [10] presented the expression

$$X_k^{\gamma,j} = \frac{(-\beta)^{|k-j|}}{(1+\beta^2)^{\gamma+1}} \sum_{s=0}^{\infty} P_s Q_s \beta^{2s} \tag{3.48}$$

where

$$P_s = \begin{cases} P_{s+k-j}, & (k \ge j) \\ P_s, & (k < j) \end{cases}$$

$$Q_s = \begin{cases} Q_j, & (k \ge j) \\ Q_{s+j-k}, & (k < j) \end{cases}$$

$$P_s = \sum_{r=0}^{s} \binom{\gamma - j + 1}{s - r} \frac{1}{r!} \left(-\frac{ke}{2\beta}\right)^r$$

$$Q_s = \sum_{r=0}^{s} \binom{\gamma + j + 1}{s - r} \frac{1}{r!} \left(\frac{ke}{2\beta}\right)^r \tag{3.49}$$

It is straightforward to prove that Eqs. (3.45), (3.48) and (3.39) are the same.

(2) Recursion Methods of the Functions in Hansen Expressions

The use of Hansen expressions to calculate Hansen coefficients requires the calculation of the functions M_s and $N_{t'}$ or generalized Laguerre polynomials. The recursion calculation methods of the functions M_s and $N_{t'}$ and the generalized Laguerre polynomials are illustrated as follows.

1. Recursion method of the functions M_s and $N_{t'}$

For the functions M_s and $N_{t'}$ in Eqs. (3.36) and (3.42), Lyddane recommended the following recursion method [17]. First, the function M_s is expressed in the form

$$M(s, \alpha; b) = (-1)^s \sum_{\nu=0}^{s} \binom{\alpha - \nu}{s - \nu} \frac{b^\nu}{\nu!} \tag{3.50}$$

Then, using Yang Hui's triangle,

$$\binom{n}{m} = \binom{n-1}{m} + \binom{n-1}{m-1} \tag{3.51}$$

Eq. (3.50) is further expressed as

$$
\begin{aligned}
M(s,\alpha;b) &= (-1)^s \sum_{\nu=0}^{s}\left[\binom{\alpha-1-\nu}{s-\nu}+\binom{\alpha-1-\nu}{s-1-\nu}\right]\frac{b^\nu}{\nu!} \\
&= M(s,\alpha-1;b)+(-1)\times(-1)^{s-1}\sum_{\nu=0}^{s-1}\binom{\alpha-1-\nu}{s-1-\nu}\frac{b^\nu}{\nu!} \\
&= M(s,\alpha-1;b)-M(s-1,\alpha-1;b) \qquad (3.52)
\end{aligned}
$$

Equation (3.52) is the recursion calculation formula for the function M_s. The function $N_{t'}$ is expressed as

$$
N_{t'} \equiv M\left(t',\gamma;-b\right) \qquad (3.53)
$$

The recursion calculation of the function $N_{t'}$ is performed using recursion formula Eq. (3.52). The functions M_s and $N_{t'}$ are recursively derived, and by substituting them into the expression for the Hansen coefficients, $X_k^{n,m}$ is calculated.

2. Recursion method of the generalized Laguerre polynomials

The generalized Laguerre polynomials $L_m^{(j)}(x)$ involved in Eqs. (3.38) and (3.39) have the recursion relation

$$
(n+1)L_{n+1}^{(\alpha)}(x)=(2n+\alpha+1-x)L_n^{(\alpha)}(x)-(n+\alpha)L_{n-1}^{(\alpha)}(x) \qquad (3.54)
$$

The corresponding initial values of the recursion are

$$
L_0^{(\alpha)}(x)=1,\quad L_1^{(\alpha)}(x)=\alpha+1-x \qquad (3.55)
$$

Adopting the generalized Laguerre polynomials for calculating Hansen coefficients, the existing Laguerre polynomials are used, and it is obviously more reasonable. This method is thus recommended.

3.1.4 Laplace Coefficient Expressions

According to the definition of the Hansen coefficient,

$$
X_k^{\gamma,m}=\frac{1}{2\pi}\int_0^{2\pi}\left(\frac{r}{a}\right)^\gamma \exp(jmv)\exp(-jkM)dM
$$

Laskar replaced the mean anomaly M by the eccentric anomaly E in the expression and obtained [20]

$$X_k^{\gamma,m} = \frac{1}{2\pi}\int_0^{2\pi}\left(\frac{r}{a}\right)^{\gamma+1}\exp(jmv)\exp(jke\sin E)\exp(-jkE)dE \tag{3.56}$$

Adopting Hansen's methodology of deducing formulae and defining $\eta = \exp(jE)$, we have

$$\frac{r}{a} = \left(1+\beta^2\right)^{-1}(1-\beta\eta)\left(1-\beta\eta^{-1}\right)$$

$$\exp(jv) = \eta\frac{1-\beta\eta^{-1}}{1-\beta\eta}$$

$$\left(\frac{r}{a}\right)^{\gamma+1}\exp(jmv) = \frac{\eta^m}{\left(1+\beta^2\right)^{\gamma+1}}(1-\beta\eta)^{\gamma+1-m}\left(1-\beta\eta^{-1}\right)^{\gamma+1+m}$$

and therefore,

$$\left(\frac{r}{a}\right)^{\gamma+1}\exp(jmv) = \frac{\eta^m}{2\left(1+\beta^2\right)^{\gamma+1}}\sum_{l=-\infty}^{\infty} b^{(l)}_{-\gamma-1+m,-\gamma-1-m}(\beta)\eta^l \tag{3.57}$$

where $b^{(k)}_{s,r}$ is the generalized Laplace coefficient defined by [21]

$$(1-\alpha z)^{-s}\left(1-\alpha z^{-1}\right)^{-r} = \frac{1}{2}\sum_{k=-\infty}^{\infty} b^{(k)}_{s,r}(\alpha)z^k \tag{3.58}$$

The generalized Laplace coefficient has the relations [20, 21] $b^{(k)}_{s,r}(-\alpha) = (-1)^k b^{(k)}_{s,r}(\alpha)$ and $b^{(-k)}_{s,r}(\alpha) = b^{(k)}_{r,s}(\alpha)$. When $k \geq 0$, $b^{(k)}_{s,r}$ can be expressed as

$$b^{(k)}_{s,r}(\alpha) = 2\frac{(s)_k}{k!}\alpha^k F\left(r, s+k, k+1; \alpha^2\right) \tag{3.59}$$

Using the generalized Laplace coefficient and the classical Bessel function expansion

$$\exp(jke\sin E) = \sum_{n=-\infty}^{\infty} J_n(ke)\eta^n \tag{3.60}$$

the Hansen coefficients are expressed as [20]

$$X_k^{\gamma,m} = \frac{1}{2\left(1+\beta^2\right)^{\gamma+1}}\sum_{n=-\infty}^{\infty} b^{(k-n-m)}_{-\gamma-1+m,-\gamma-1-m}(\beta)J_n(ke) \tag{3.61}$$

In addition, Laskar [20] presented five recursion relations of the generalized Laplace coefficients:

$$\begin{aligned}
&b_{s,r+1}^{(k)}(\alpha) - \alpha b_{s,r+1}^{(k+1)}(\alpha) = b_{s,r}^{(k)}(\alpha) \\
&b_{s+1,r}^{(k)}(\alpha) - \alpha b_{s+1,r}^{(k-1)}(\alpha) = b_{s,r}^{(k)}(\alpha) \\
&\left(1-\alpha^2\right) r b_{s,r+1}^{(k)}(\alpha) = (r-k) b_{s,r}^{(k)}(\alpha) + \alpha(s+k-1) b_{s,r}^{(k-1)}(\alpha) \\
&\left(1-\alpha^2\right) s b_{s+1,r}^{(k)}(\alpha) = (s+k) b_{s,r}^{(k)}(\alpha) + \alpha(r-k-1) b_{s,r}^{(k+1)}(\alpha) \\
&(r-k-1) b_{s,r}^{(k+1)}(\alpha) = (s+k-1) b_{s,r}^{(k-1)}(\alpha) + \left[\alpha(r-s-k) - \frac{k}{\alpha}\right] b_{s,r}^{(k)}(\alpha)
\end{aligned} \tag{3.62}$$

and pointed out that by using these five recursion relations, all the generalized Laplace coefficients can be expressed as functions of the two coefficients $b_{s-|s|,r-|r|}^{(0)}(\alpha)$ and $b_{s-|s|,r-|r|}^{(1)}(\alpha)$. In particular, all the generalized coefficients in the expression of Hansen coefficients Eq. (3.61) can be expressed as functions of the two Laplace coefficients $b_{|\gamma|-\gamma}^{(0)}(\alpha)$ and $b_{|\gamma|-\gamma}^{(1)}(\alpha)$.

When $k-n-m \geq 0$, using Eq. (3.59), it is not difficult to prove that

$$\begin{aligned}
b_{-\gamma-1+m,-\gamma-1-m}^{(k-n-m)} &= 2\beta^{k-n-m} \sum_{s=0}^{\infty} \binom{-\gamma-2-m+s}{s} \binom{-\gamma-2+k-n+s}{k-n-m+s} \beta^{2s} \\
&= 2(-\beta)^{k-n-m} \sum_{s=0}^{\infty} \binom{\gamma+m+1}{s} \binom{\gamma-m+1}{k-n-m+s} \beta^{2s}
\end{aligned}$$

Equation (3.61) can then be rewritten as

$$\begin{aligned}
X_k^{\gamma,m} = &\frac{1}{\left(1+\beta^2\right)^{\gamma+1}} \sum_{n=-\infty}^{\infty} (-\beta)^{k-n-m} \sum_{s=0}^{\infty} \binom{\gamma+m+1}{s} \binom{\gamma-m+1}{k-n-m+s} \beta^{2s} \\
&\times J_n(ke)
\end{aligned} \tag{3.63}$$

Thus, the Laplace coefficient expression is equivalent to Wnuk's expression Eq. (3.17). However, it also provides a means for calculating Hansen coefficients; i.e., as long as the generalized Laplace coefficient and Bessel function are calculated, the Hansen coefficients can be immediately obtained.

Lion and Métris [22] proposed a recursion method of the generalized Laplace coefficients when studying Hansen-like coefficients of an eccentric anomaly. The proposed method has no singularity in the recursion calculation, and it is very stable and highly accurate. However, this method is not applicable to the calculation of $b_{-\gamma-1+m,-\gamma-1-m}^{(k-n-m)}$.

All the recursion relations in Eq. (3.62) have singularities for the recursion with respect to the superscript k. It remains necessary to further study methods of calculating the generalized Laplace coefficients in the calculation of Hansen coefficients of the mean anomaly. Before a suitable recursion calculation method is developed, the Laplace coefficient method can only be restored to Wnuk's method [12].

3.1.5 Definite Integral Expressions

The Hansen coefficient $X_k^{n,m}$ is defined as

$$\left(\frac{r}{a}\right)^n \exp(jmf) = \sum_{k=-\infty}^{\infty} X_k^{n,m} \exp(jkM)$$

According to the orthogonality of trigonometric functions, the definite integral expression of Hansen coefficients $X_k^{n,m}$ is

$$X_k^{n,m} = \frac{1}{2\pi} \int_0^{2\pi} \left(\frac{r}{a}\right)^n \cos(mf - kM) dM \tag{3.64}$$

The integral variable of the definite integral expression Eq. (3.64) is the mean anomaly M. However, a trial calculation proved that higher accuracy can be achieved using the eccentric anomaly E as the integral variable. In this way, the definite integral expression of Hansen coefficients is

$$X_k^{n,m} = \frac{1}{\pi} \int_0^{\pi} \left(\frac{r}{a}\right)^{n+1} \cos(mf - kM) dE \tag{3.65}$$

Adopting Eqs. (3.64) and (3.65), Hansen coefficients can be calculated using the definite integral method. This is the most direct method of calculating Hansen coefficients and is easy to understand. The definite integral method can guarantee calculation accuracy but not the number of significant digits in the calculation results. Using the symmetry of Hansen coefficients $X_k^{n,m} = X_{-k}^{n,-m}$(or $G_{lpq} = G_{l,l-p,-q}$), the cost of calculating Hansen coefficients with the definite integral method can be reduced by nearly half.

3.2 Kernels of Hansen Coefficients

The kernels $K_t^{n,s}$ of Hansen coefficients are needed in the perturbation theory without singularities and are defined as

$$K_t^{n,s} = X_t^{n,s} e^{-|t-s|} \tag{3.66}$$

As there is no factor $e^{|t-s|}$ in the Hill expressions of Hansen coefficients, it is inconvenient to calculate $K_t^{n,s}$ using Hill expressions. Therefore, only Eqs. (3.38), (3.39) and (3.10) are available to calculate $K_t^{n,s}$.

According to Eq. (3.38), we have

$$K_t^{n,s} = A \sum_{i=0}^{\infty} L_{i+a}^{(n-t+1)}(\mu) L_{i+b}^{(n+t+1)}(-\mu) \beta^{2i} \tag{3.67}$$

where

$$A = \left(1-\beta^2\right)^{2n+3} \left(1+\beta^2\right)^{-(n+1)} \left(-\frac{1+\beta^2}{2}\right)^{|t-s|} \tag{3.68}$$

The calculation results are just the Hansen coefficients $X_t^{n,s}$, if A in the expression Eq. (3.67) of the kernels $K_t^{n,s}$ of Hansen coefficients is replaced with $A = \left(1-\beta^2\right)^{2n+3}\left(1+\beta^2\right)^{-(n+1)}(-\beta)^{|t-s|}$.

Similarly, according to Eq. (3.39), we have

$$K_t^{n,s} = B \sum_{i=0}^{\infty} L_{i+a}^{(n-s-i-a+1)}(\nu) L_{i+b}^{(n+s-i-b+1)}(-\nu) \beta^{2i} \tag{3.69}$$

where

$$B = \left(1+\beta^2\right)^{-(n+1)} \left(-\frac{1+\beta^2}{2}\right)^{|t-s|} \tag{3.70}$$

Similarly, if B in the expression Eq. (3.69) of the kernels $K_t^{n,s}$ of Hansen coefficients is replaced with $B = \left(1+\beta^2\right)^{-(n+1)}(-\beta)^{|t-s|}$, the calculation results are just the Hansen coefficients $X_t^{n,s}$. The methods of calculating the kernels $K_t^{n,s}$ of Hansen coefficients using Eqs. (3.67) and (3.69) are the same as the methods of calculating Hansen coefficients using Eqs. (3.38) and (3.39). Correspondingly, using Eqs. (3.67) and (3.69), the derivatives of the kernels of Hansen coefficients are calculated as

$$\begin{aligned} \frac{dK_t^{n,s}}{de} &= 2e\frac{dK_t^{n,s}}{de^2} \\ \frac{dK_t^{n,s}}{de^2} &= \frac{dK_t^{n,s}}{d\beta^2}\frac{d\beta^2}{de^2} = \frac{\left(1+\beta^2\right)^3}{4\left(1-\beta^2\right)}\frac{dK_t^{n,s}}{d\beta^2} \end{aligned} \tag{3.71}$$

According to Eq. (3.10), we have

$$K_{m+s}^{n,m} = \left(-\frac{1}{2}\right)^s$$

$$\times \sum_{t=0}^{\infty} \left(\frac{e}{2}\right)^{2t} \left\{ \sum_{j=0}^{t} \sum_{p=0}^{j} \binom{n+m+1}{j-p} X_p \sum_{q=0}^{s+j} \binom{n-m+1}{s+j-q} X_q \left[\binom{k}{t-j} - \binom{k}{t-j-1} \right] \right\}$$

Of course, if $s < 0$, then the indexes m and s in the above expression change signs simultaneously. The derivatives of the kernels of Hansen coefficients are derived by directly differentiating the above expression with respect to e.

3.3 Direct Methods of Calculating the Derivatives of Hansen Coefficients

3.3.1 *Derivative Calculations Using the Definite Integral Method*

By differentiating Eq. (3.64) with respect to eccentricity e, the definite integral expression of the derivatives of Hansen coefficients using the mean anomaly M as the integral variable is immediately obtained as

$$\begin{aligned} \frac{dX_k^{n,m}}{de} = \frac{1}{2\pi} \int_0^{2\pi} -\left(\frac{r}{a}\right)^n \Big[& n\frac{a}{r} \cos f \cos(mf - kM) \\ & + \frac{m}{1-e^2} (2 + e\cos f) \sin f \sin(mf - kM) \Big] dM \end{aligned} \tag{3.72}$$

By differentiating Eq. (3.65) with respect to eccentricity e, the definite integral expression of the derivatives of Hansen coefficients using the eccentric anomaly E as the integral variable is immediately obtained as

$$\begin{aligned} \frac{dX_k^{n,m}}{de} = \frac{1}{\pi} \int_0^{\pi} -\left(\frac{r}{a}\right)^{n+1} \Big[& n\frac{a}{r} \cos f \cos(mf - kM) \\ & + \frac{m}{1-e^2} (2 + e\cos f) \sin f \sin(mf - kM) \Big] dE \end{aligned} \tag{3.73}$$

3.3.2 *Derivative Calculations Using Hansen Coefficients*

Challe and Laclaverie presented a pair of formulae for calculating the derivatives of Hansen coefficients [23] (see Appendix A.2 for the specific deduction):

$$\begin{aligned} 2e\left(1-e^2\right)\frac{dX_k^{n,m}}{de} = & \left[2m + (m-n)e^2 - 2k\left(1-e^2\right)^{3/2}\right] X_k^{n,m} \\ & + (4m - 2n)e X_k^{n,m+1} + (m-n)e^2 X_k^{n,m+2} \end{aligned} \tag{3.74a}$$

$$2e\left(1-e^2\right)\frac{dX_k^{n,m}}{de}=-\left[2m+(n+m)e^2-2k\left(1-e^2\right)^{3/2}\right]X_k^{n,m}\\-(2n+4m)eX_k^{n,m-1}-(n+m)e^2X_k^{n,m-2}\qquad(3.74\text{b})$$

Wnuk suggested using Eq. (3.74b) to calculate the derivatives of Hansen coefficients [12]. This formula seems to have a singularity of $e = 0$, but it can be transformed so as not to have a singularity. In fact, Eq. (3.74b) can be rewritten as

$$2\left(1-e^2\right)\frac{dX_k^{n,m}}{de}=-\frac{2(m-k)}{e}X_k^{n,m}-(n+m)eX_k^{n,m}-2k\frac{1-\left(1-e^2\right)^{3/2}}{e}X_k^{n,m}\\-(2n+4m)X_k^{n,m-1}-(n+m)eX_k^{n,m-2}$$

As there is a factor of e in $X_k^{n,m}$ when $m-k\neq 0$, and

$$\frac{1-\left(1-e^2\right)^{3/2}}{e}=\frac{1-\left(1-e^2\right)^{1/2}+\left(1-e^2\right)^{1/2}\left[1-\left(1-e^2\right)\right]}{e}\\=\beta+e\left(1-e^2\right)^{1/2}$$

there is no denominator of e in Eq. (3.74b) in fact, but it needs to be distinguished many times in the calculation. Another problem with Eq. (3.74b) is that $X_k^{n,m-1}$ in the formula cannot be transformed into an eccentricity function. However, the calculation cost of this method is low, and only two Hansen coefficients need to be calculated additionally.

The denominator of e can be avoided by adopting the formula given by Giacaglia [13]:

$$\frac{dX_k^{n,m}}{de}=\frac{2m-n}{2}X_k^{n-1,m+1}-\frac{n+2m}{2}X_k^{n-1,m-1}-\frac{me}{4\left(1-e^2\right)}\left(X_k^{n,m+2}-X_k^{n,m-2}\right)$$

$$\frac{dG_{lpq}}{de}=\frac{3l-4p+1}{2}G_{l+1,p,q-1}-\frac{l-4p-1}{2}G_{l+1,p+1,q+1}\\-\frac{(l-2p)e}{4\left(1-e^2\right)}\left(G_{l,p-1,q-2}-G_{l,p+1,q+2}\right)\qquad(3.75)$$

However, this increases the calculation cost, and four Hansen coefficients need to be calculated to obtain a derivative.

3.3.3 Direct Derivation Methods

Adopting the expressions of the Hansen coefficients given in Eqs. (3.10), (3.17), (3.23), (3.38) and (3.39) and by directly differentiating Hansen coefficients with respect to the eccentricity e, the calculation formulae of the derivatives of Hansen coefficients can be obtained. These direct derivation formulae are introduced in detail below.

(1) Direct Derivation of Eq. (3.38)

According to Eq. (3.38), we have

$$X_t^{n,s} = AK\left(-\frac{e}{2}\right)^{|t-s|} \tag{3.76}$$

where

$$K = \sum_{i=0}^{\infty} L_{i+a}^{(n-t+1)}(\mu) L_{i+b}^{(n+t+1)}(-\mu)\beta^{2i} \tag{3.77}$$

$$A = \left(1-\beta^2\right)^{2n+3}\left(1+\beta^2\right)^{-(n+1)+|t-s|}, \quad \mu = t\left(1-\beta^2\right)/\left(1+\beta^2\right) \tag{3.78}$$

We use the formula for calculating the derivative of the generalized Laguerre polynomials,

$$\frac{dL_n^{(\alpha)}(x)}{dx} = -L_{n-1}^{(\alpha+1)}(x)$$

and define

$$\begin{aligned}\frac{dLL}{d\mu} &\equiv \frac{d\left[L_{i+a}^{(n-t+1)}(\mu)L_{i+b}^{(n+t+1)}(-\mu)\right]}{d\mu} \\ &= L_{i+a}^{(n-t+1)}(\mu)L_{i+b-1}^{(n+t+2)}(-\mu) - L_{i+a-1}^{(n-t+2)}(\mu)L_{i+b}^{(n+t+1)}(-\mu)\end{aligned}$$

By directly differentiating the Hansen coefficients in Eq. (3.76) with respect to e, the formula for calculating the derivatives of Hansen coefficients is derived immediately as

$$\frac{dX_t^{n,s}}{de} = \left(\frac{dA}{d\beta^2}K + A\frac{dK}{d\beta^2}\right)\left(-\frac{e}{2}\right)^{|t-s|}\frac{d\beta^2}{de} + AK\left(-\frac{1}{2}\right)^{|t-s|}|t-s|e^{|t-s|-1} \tag{3.79}$$

where

$$\frac{dA}{d\beta^2} = A\left[\frac{-(2n+3)}{1-\beta^2} + \frac{-(n+1)+|t-s|}{1+\beta^2}\right]$$

$$\frac{dK}{d\beta^2} = \sum_{i=0}^{\infty}\left[\frac{dLL}{d\mu}\frac{d\mu}{d\beta^2}\beta^{2i} + iL_{i+a}^{(n-t+1)}(\mu)L_{i+b}^{(n+t+1)}(-\mu)\beta^{2i-2}\right]$$
$$\frac{d\beta^2}{de} = \frac{e(1+\beta^2)^3}{2(1-\beta^2)}$$
$$\frac{d\mu}{d\beta^2} = -2t(1+\beta^2)^{-2} \tag{3.80}$$

(2) Direct Derivation of Eq. (3.39)

According to Eq. (3.39), we have

$$X_t^{n,s} = AK\left(-\frac{e}{2}\right)^{|t-s|} \tag{3.81}$$

where

$$K = \sum_{i=0}^{\infty} L_{i+a}^{(n-s-i-a+1)}(\nu)L_{i+b}^{(n+s-i-b+1)}(-\nu)\beta^{2i} \tag{3.82}$$

$$A = (1+\beta^2)^{-(n+1)+|t-s|}, \quad \nu = t/(1+\beta^2) \tag{3.83}$$

We define

$$\frac{dLL}{d\nu} \equiv \frac{d\left[L_{i+a}^{(n-s-i-a+1)}(\nu)L_{i+b}^{(n+s-i-b+1)}(-\nu)\right]}{d\nu}$$
$$= L_{i+a}^{(n-s-i-a+1)}(\nu)L_{i+b-1}^{(n+s-i-b+2)}(-\nu) - L_{i+a-1}^{(n-s-i-a+2)}(\nu)L_{i+b}^{(n+s-i-b+1)}(-\nu)$$

By directly differentiating the Hansen coefficients in Eq. (3.81) with respect to e, the formula for calculating the derivatives of Hansen coefficients is derived immediately as

$$\frac{dX_t^{n,s}}{de} = \left(\frac{dA}{d\beta^2}K + A\frac{dK}{d\beta^2}\right)\left(-\frac{e}{2}\right)^{|t-s|}\frac{d\beta^2}{de} + AK\left(-\frac{1}{2}\right)^{|t-s|}|t-s|e^{|t-s|-1} \tag{3.84}$$

where

$$\frac{dA}{d\beta^2} = A\left[\frac{-(n+1)+|t-s|}{1+\beta^2}\right]$$
$$\frac{dK}{d\beta^2} = \sum_{i=0}^{\infty}\left[\frac{dLL}{d\nu}\frac{d\nu}{d\beta^2}\beta^{2i} + iL_{i+a}^{(n-s-i-a+1)}(\nu)L_{i+b}^{(n+s-i-b+1)}(-\nu)\beta^{2i-2}\right]$$

$$\frac{d\beta^2}{de} = \frac{e\left(1+\beta^2\right)^3}{2\left(1-\beta^2\right)}$$
$$\frac{dv}{d\beta^2} = -t\Big/\left(1+\beta^2\right)^2 \tag{3.85}$$

(3) Direct Derivation of Eq. (3.10)

By directly differentiating the Hansen coefficients in Eq. (3.10) with respect to e, the expression for calculating the derivatives of Hansen coefficients is derived as

$$\frac{dX_{m+s}^{n,m}}{de} = (-1)^s \sum_{t=0}^{\infty} \left\{ \sum_{j=0}^{t} \sum_{p=0}^{j} \binom{n+m+1}{j-p} X_p \sum_{q=0}^{s+j} \binom{n-m+1}{s+j-q} X_q \right.$$
$$\left. \times \left[\binom{k}{t-j} - \binom{k}{t-j-1} \right] \right\} \frac{2t+s}{2} \left(\frac{e}{2}\right)^{2t+s-1} \tag{3.86}$$

Of course, it is noted that when the index $s < 0$, both the indexes m and s in the above formula change signs. Refer to Eq. (3.10) for other symbols in the formula.

(4) Direct Derivation of Eq. (3.17)

The expression Eq. (3.17) for calculating the Hansen coefficients can be rewritten as

$$X_k^{n,m} = \sum_{t=-\infty}^{\infty} \sum_{s=0}^{\infty} \binom{n \mp m+1}{|k-t-m|+s} \binom{n \pm m+1}{s} A(\beta) J_t(ke) \tag{3.87}$$

where

$$A(\beta) = \left(1+\beta^2\right)^{-(n+1)} (-\beta)^{|k-t-m|} \beta^{2s}$$

The plus or minus sign of the index m in Eq. (3.87) depends on whether the value of $k - t - m$ is > 0.

Then, by directly differentiating the Hansen coefficients in Eq. (3.87) with respect to e, the expression for calculating the derivatives of Hansen coefficients is derived as

$$\frac{dX_k^{n,m}}{de} = \sum_{t=-\infty}^{\infty} \sum_{s=0}^{\infty} \binom{n \mp m+1}{|k-t-m|+s} \binom{n \pm m+1}{s}$$
$$\times \left[\frac{dA(\beta)}{d\beta} \frac{d\beta}{de} J_t(ke) + A(\beta) k \frac{dJ_t(ke)}{d(ke)} \right] \tag{3.88}$$
$$\frac{dA(\beta)}{d\beta} = A(\beta) \left[\frac{-2\beta(n+1)}{1+\beta^2} + \frac{2s+|k-t-m|}{\beta} \right]$$

$$
\begin{aligned}
\frac{d\beta}{de} &= \frac{\left(1+\beta^2\right)^2}{2\left(1-\beta^2\right)} \\
\frac{dJ_t(ke)}{d(ke)} &= \frac{1}{2}\left[J_{t-1}(ke) - J_{t+1}(ke)\right]
\end{aligned} \tag{3.89}
$$

(5) Direct Derivation of Eq. (3.23)

The expression Eq. (3.23) for calculating the eccentricity functions can be rewritten as

$$
G_{lpq} = \sum_{s=0}^{s_1}\sum_{t=0}^{t_1}\binom{2l-2p+s-1}{s}\binom{2p+t-1}{t}A(\beta)J_{q-s+t}(ke) \tag{3.90}
$$

where

$$
A(\beta) = \left(1+\beta^2\right)^l \beta^{s+t}
$$

Then, by directly differentiating the eccentricity functions in Eq. (3.90) with respect to e, the expression for calculating the derivatives of eccentricity functions is derived as

$$
\begin{aligned}
\frac{dG_{lpq}}{de} = &\sum_{s=0}^{s_1}\sum_{t=0}^{t_1}\binom{2l-2p+s-1}{s}\binom{2p+t-1}{t} \\
&\times\left[\frac{dA(\beta)}{d\beta}\frac{d\beta}{de}J_{q-s+t}(ke) + A(\beta)k\frac{dJ_{q-s+t}(ke)}{d(ke)}\right]
\end{aligned} \tag{3.91}
$$

where

$$
\begin{aligned}
\frac{dA(\beta)}{d\beta} &= A(\beta)\left(\frac{2l\beta}{1+\beta^2} + \frac{t+s}{\beta}\right) \\
\frac{d\beta}{de} &= \frac{\left(1+\beta^2\right)^2}{2\left(1-\beta^2\right)} \\
\frac{dJ_{q-s+t}(ke)}{d(ke)} &= \frac{1}{2}\left[J_{q-s+t-1}(ke) - J_{q-s+t+1}(ke)\right]
\end{aligned} \tag{3.92}
$$

3.4 Trial Calculations

Various direct methods of calculating Hansen coefficients were reviewed above, and the corresponding calculation expressions were presented. Many of the expressions are equivalent, such that only the following six expressions are independent.

(1) Balmino's expression Eq. (3.10). The corresponding direct method of calculating Hansen coefficients is called Balmino's method.
(2) Two Hill expressions, namely Wnuk's expression Eq. (3.17) and Giacaglia's expression Eq. (3.23). The corresponding direct methods of calculating Hansen coefficients are called Wnuk's method and Giacaglia's method.
(3) Two Hansen expressions, namely McClain's expressions Eqs. (3.38) and (3.39). The corresponding direct methods of calculating Hansen coefficients are called McClain's first and second methods.
(4) Definite integral expression Eq. (3.65). The corresponding direct method of calculating Hansen coefficients is called the definite integral method.

It is noted that the symbols of the summation indexes in the corresponding mathematical expressions match those in the original references so that readers can consult the original texts.

Fortran programs have been compiled to compare these direct methods of calculating Hansen coefficients and their derivatives (see Appendix D.2).

(1) Balmino's method, i.e., Eqs. (3.10) and (3.86), where the summation range of the index t is $0 \sim 50$.
(2) Wnuk's method, i.e., Eqs. (3.17) and (3.74b), or Eqs. (3.87) and (3.74b), where the summation ranges of the indexes s and t are $0 \sim 70$ and $-70 \sim 70$, respectively.
(3) Wnuk's direct derivation method, i.e., Eqs. (3.17) and (3.88), or Eqs. (3.87) and (3.88), where the summation ranges of the indexes s and t are respectively $0 \sim 70$ and $-70 \sim 70$ and we note that the expressions for calculating Hansen coefficients in Wnuk's method and Wnuk's direct derivation method are the same but the expressions for calculating the derivatives of Hansen coefficients are different.
(4) Giacaglia's method, i.e., Eqs. (3.23) and (3.91), or Eqs. (3.90) and (3.91), where the summation range of the index s is $0 \sim 70$ when the index $l \neq p$ and the summation range of the index t is $0 \sim 70$ when the index $p \neq 0$.
(5) McClain's first method, i.e., Eqs. (3.38) and (3.79), or Eqs. (3.76) and (3.79), where the summation range of the index i is $0 \sim 70$.
(6) McClain's second method, i.e., Eqs. (3.39) and (3.84), or Eqs. (3.81) and (3.84), where the summation range of the index i is $0 \sim 70$.
(7) Definite integral method, i.e., Eqs. (3.65) and (3.73), where the two-point Gauss integral method is adopted in 2000 sections.

Wnuk made concise remarks on the corresponding direct methods of calculating Hansen coefficients and their derivatives [12].

(1) Wnuk's method with the modified Bessel function (asymptotic expansion) algorithm is digitally stable and is based on the mathematical expressions in Eqs. (3.17) and (3.74b). In addition, it performs fast computations of eccentricity functions $G_{lpq} = X_{l-2p+q}^{-(l+1),l-2p}$ and their derivatives for eccentricities $e < 1$ and higher-order l, p and q.
(2) Kaula's method [18], i.e., Eq. (3.45), is digitally unstable for large eccentricities and essentially has the same mathematical expression as Eq. (3.39) of McClain's second method.
(3) The definite integral method, i.e., Eq. (3.65), has good computational performance for large eccentricities but limited practical application to eccentricity functions with higher-order l, p and q owing to its computing time and digital instability.

3.4.1 Calculation Results for Hansen Coefficients

Trail calculations were carried out using the compiled Fortran programs of the direct methods of calculating Hansen coefficients and their derivatives. The calculation results are given in Tables 3.3, 3.4 and 3.5.

Table 3.3 shows that the calculation results of the various methods are in good agreement with each other. Moreover, for small eccentricities, the various direct methods of calculating Hansen coefficients meet the requirements of accuracy, and there are no issues in implementing the presented calculation formulae and the complied Fortran programs in this monograph.

As stated above, Wnuk [12] pointed out that Kaula's method, i.e., Eq. (3.45), is unstable for large eccentricities. Because Eq. (3.45) of Kaula's method is equivalent to Eq. (3.39) or Eq. (3.81) of McClain's second method, in Wnuk's [12] opinion, McClain's second method is also unstable. However, as seen from Table 3.4, the calculation results of Balmino's method, McClain's first method, and McClain's second method, i.e., Eqs. (3.10), (3.38) and (3.39), or Eqs. (3.10), (3.76) and (3.81), are good as long as the upper limits of the summation of the corresponding indexes are large enough. Therefore, in the calculations of lower-order Hansen coefficients (i.e., the index k is relatively small), the various direct methods also satisfy the requirement of calculation accuracy for large eccentricities. Although some of the methods converge slowly, there is no computational instability.

Wnuk [12] pointed out that Wnuk's method with the modified Bessel function (asymptotic expansion) algorithm, i.e., Eq. (3.17) or Eq. (3.87), is digitally stable but did not examine the computational stabilities of Giacaglia's method, McClain's first method and McClain's second method, i.e., Eqs. (3.23), (3.38) and (3.39), or Eqs. (3.90), (3.76) and (3.81). To clarify whether these methods are stable, we calculate Hansen coefficients for the example that $e = 0.6$. The calculation results are listed in Table 3.5.

Table 3.5 shows that the results of Wnuk's method and Giacaglia's method, i.e., Eqs. (3.17) and (3.23), or Eqs. (3.87) and (3.90), are in good agreement and that the

Table 3.3 Calculation example of Hansen coefficients ($e = 0.1$)

l	p	q	k	Balmino's method Eq. (3.10)	Wnuk's method Eq. (3.17) or Eq. (3.87)	Giacaglia's method Eq. (3.23) or Eq. (3.90)
2	0	−6	−4	8.933773426754E−08	8.933773426754E−08	8.933773426754E−08
2	0	−5	−3	6.368016123798E−07	6.368016123798E−07	6.368016123798E−07
2	0	−4	−2	4.196062556964E−06	4.196062556964E−06	4.196062556964E−06
2	0	−3	−1	2.097758902378E−05	2.097758902378E−05	2.097758902378E−05
2	0	−2	0	0.000000000000E+00	0.000000000000E+00	0.000000000000E+00
2	0	−1	1	−4.993763099038E−02	−4.993763099038E−02	−4.993763099038E−02
2	0	0	2	9.750811283840E−01	9.750811283840E−01	9.750811283840E−01
2	0	1	3	3.423506171233E−01	3.423506171233E−01	3.423506171233E−01
2	0	2	4	8.309581470911E−02	8.309581470911E−02	8.309581470911E−02
2	0	3	5	1.718403933461E−02	1.718403933461E−02	1.718403933461E−02
2	0	4	6	3.245640680805E−03	3.245640680805E−03	3.245640680805E−03
2	0	5	7	5.781706326503E−04	5.781706326503E−04	5.781706326503E−04
2	0	6	8	9.886019079453E−05	9.886019079453E−05	9.886019079453E−05
2	1	−6	−6	9.858368157687E−06	9.858368157687E−06	9.858368157687E−06
2	1	−5	−5	6.918031769307E−05	6.918031769307E−05	6.918031769307E−05
2	1	−4	−4	4.820846127921E−04	4.820846127921E−04	4.820846127921E−04
2	1	−3	−3	3.328095902056E−03	3.328095902056E−03	3.328095902056E−03
2	1	−2	−2	2.267722802380E−02	2.267722802380E−02	2.267722802380E−02
2	1	−1	−1	1.517081261347E−01	1.517081261347E−01	1.517081261347E−01
2	1	0	0	1.015189712383E+00	1.015189712383E+00	1.015189712383E+00
2	1	1	1	1.517081261347E−01	1.517081261347E−01	1.517081261347E−01
2	1	2	2	2.267722802380E−02	2.267722802380E−02	2.267722802380E−02
2	1	3	3	3.328095902056E−03	3.328095902056E−03	3.328095902056E−03
2	1	4	4	4.820846127921E−04	4.820846127921E−04	4.820846127921E−04
2	1	5	5	6.918031769307E−05	6.918031769307E−05	6.918031769307E−05
2	1	6	6	9.858368157687E−06	9.858368157687E−06	9.858368157687E−06
l	p	q	k	McClain's first method Eq. (3.38) or Eq. (3.76)	McClain's second method Eq. (3.39) or Eq. (3.81)	Definite integral method Eq. (3.65)
2	0	−6	−4	8.933773426754E−08	8.933773426754E−08	8.933773434533E−08
2	0	−5	−3	6.368016123798E−07	6.368016123798E−07	6.368016124214E−07
2	0	−4	−2	4.196062556964E−06	4.196062556964E−06	4.196062556998E−06
2	0	−3	−1	2.097758902378E−05	2.097758902378E−05	2.097758902332E−05
2	0	−2	0	0.000000000000E+00	1.114890262054E−135	−4.259925745487E−16
2	0	−1	1	−4.993763099038E−02	−4.993763099038E−02	−4.993763099038E−02
2	0	0	2	9.750811283840E−01	9.750811283840E−01	9.750811283840E−01
2	0	1	3	3.423506171233E−01	3.423506171233E−01	3.423506171233E−01

(continued)

Table 3.3 (continued)

l	p	q	k	McClain's first method Eq. (3.38) or Eq. (3.76)	McClain's second method Eq. (3.39) or Eq. (3.81)	Definite integral method Eq. (3.65)
2	0	2	4	8.309581470911E−02	8.309581470911E−02	8.309581470912E−02
2	0	3	5	1.718403933461E−02	1.718403933461E−02	1.718403933461E−02
2	0	4	6	3.245640680805E−03	3.245640680805E−03	3.245640680805E−03
2	0	5	7	5.781706326503E−04	5.781706326503E−04	5.781706326504E−04
2	0	6	8	9.886019079453E−05	9.886019079453E−05	9.886019079444E−05
2	1	−6	−6	9.858368157687E−06	9.858368157687E−06	9.858368157794E−06
2	1	−5	−5	6.918031769307E−05	6.918031769307E−05	6.918031769320E−05
2	1	−4	−4	4.820846127921E−04	4.820846127921E−04	4.820846127920E−04
2	1	−3	−3	3.328095902056E−03	3.328095902056E−03	3.328095902056E−03
2	1	−2	−2	2.267722802380E−02	2.267722802380E−02	2.267722802380E−02
2	1	−1	−1	1.517081261347E−01	1.517081261347E−01	1.517081261347E−01
2	1	0	0	1.015189712383E+00	1.015189712383E+00	1.015189712383E+00
2	1	1	1	1.517081261347E−01	1.517081261347E−01	1.517081261347E−01
2	1	2	2	2.267722802380E−02	2.267722802380E−02	2.267722802380E−02
2	1	3	3	3.328095902056E−03	3.328095902056E−03	3.328095902056E−03
2	1	4	4	4.820846127921E−04	4.820846127921E−04	4.820846127920E−04
2	1	5	5	6.918031769307E−05	6.918031769307E−05	6.918031769320E−05
2	1	6	6	9.858368157687E−06	9.858368157687E−06	9.858368157794E−06

results of the definite integral method, i.e., Eq. (3.65), are consistent with the results of Wnuk's method and Giacaglia's method on the whole. However, the results of Balmino's method and McClain's first method, i.e., Eqs. (3.10) and (3.38), or Eqs. (3.10) and (3.76), do not agree with the results of Wnuk's method from $q \geq -4$, and the results of McClain's second method, i.e., Eq. (3.39) or Eq. (3.81), do not agree with the results of Wnuk's method from $q \geq 6$. The calculation results indicate instabilities of these methods in the calculation of higher-order Hansen coefficients (i.e., the index k is relatively large) at large eccentricities. The root of these instabilities is the insufficient word length of the processing computer. If quadruple precision computation is adopted, the results become stable.

As stated above, Wnuk's method and Wnuk's direct derivation method have the same calculation expressions for Hansen coefficients. The corresponding Hansen coefficients calculated with these two methods are obviously the same.

Table 3.4 Calculation example of Hansen coefficients ($e = 0.75$)

l	p	q	k	Balmino's method Eq. (3.10)	Wnuk's method Eq. (3.17) or Eq. (3.87)	Giacaglia's method Eq. (3.23) or Eq. (3.90)
2	0	0	2	−1.731234895046E−01	−1.731234895046E−01	−1.731234895046E−01
2	0	1	3	1.777658131087E−01	1.777658131087E−01	1.777658131087E−01
2	0	2	4	5.848818366188E−01	5.848818366188E−01	5.848818366188E−01
2	0	3	5	9.915543956460E−01	9.915543956460E−01	9.915543956460E−01
2	0	4	6	1.367580969223E+00	1.367580969223E+00	1.367580969223E+00
2	0	5	7	1.697429520264E+00	1.697429520264E+00	1.697429520264E+00
2	0	6	8	1.974320850707E+00	1.974320850707E+00	1.974320850707E+00
2	0	7	9	2.196833454866E+00	2.196833454866E+00	2.196833454866E+00
2	0	8	10	2.366808728706E+00	2.366808728706E+00	2.366808728706E+00
2	0	9	11	2.488005529897E+00	2.488005529897E+00	2.488005529897E+00
2	0	10	12	2.565218400475E+00	2.565218400474E+00	2.565218400474E+00
2	0	11	13	2.603696309314E+00	2.603696309315E+00	2.603696309315E+00
2	0	12	14	2.608762137585E+00	2.608762137586E+00	2.608762137586E+00
2	0	13	15	2.585568795367E+00	2.585568795368E+00	2.585568795368E+00
2	0	14	16	2.538949276973E+00	2.538949276974E+00	2.538949276974E+00
2	0	15	17	2.473331486938E+00	2.473331486944E+00	2.473331486944E+00
2	0	16	18	2.392697545777E+00	2.392697545767E+00	2.392697545767E+00

l	p	q	k	McClain's first method Eq. (3.38) or Eq. (3.76)	McClain's second method Eq. (3.39) or Eq. (3.81)	Definite integral method Eq. (3.65)
2	0	0	2	−1.731234895046E−01	−1.731234895046E−01	−1.731234895046E−01
2	0	1	3	1.777658131087E−01	1.777658131087E−01	1.777658131087E−01
2	0	2	4	5.848818366188E−01	5.848818366188E−01	5.848818366188E−01
2	0	3	5	9.915543956460E−01	9.915543956460E−01	9.915543956460E−01
2	0	4	6	1.367580969223E+00	1.367580969223E+00	1.367580969223E+00
2	0	5	7	1.697429520264E+00	1.697429520264E+00	1.697429520264E+00
2	0	6	8	1.974320850707E+00	1.974320850707E+00	1.974320850707E+00
2	0	7	9	2.196833454866E+00	2.196833454866E+00	2.196833454866E+00
2	0	8	10	2.366808728706E+00	2.366808728706E+00	2.366808728706E+00
2	0	9	11	2.488005529897E+00	2.488005529897E+00	2.488005529897E+00
2	0	10	12	2.565218400475E+00	2.565218400474E+00	2.565218400475E+00
2	0	11	13	2.603696309314E+00	2.603696309314E+00	2.603696309314E+00
2	0	12	14	2.608762137586E+00	2.608762137585E+00	2.608762137586E+00
2	0	13	15	2.585568795367E+00	2.585568795368E+00	2.585568795367E+00
2	0	14	16	2.538949276973E+00	2.538949276972E+00	2.538949276973E+00
2	0	15	17	2.473331486938E+00	2.473331486935E+00	2.473331486938E+00
2	0	16	18	2.392697545777E+00	2.392697545774E+00	2.392697545777E+00

Table 3.5 Calculation example of Hansen coefficients ($e = 0.6$)

l	p	q	k	Balmino's method Eq. (3.10)	Wnuk's method Eq. (3.17) or Eq. (3.87)	Giacaglia's method Eq. (3.23) or Eq. (3.90)
30	2	−20	6	2.933226736230E−04	2.933228305423E−04	2.933228305424E−04
30	2	−18	8	5.309383355001E−04	5.309386709805E−04	5.309386709814E−04
30	2	−16	10	9.610462350119E−04	9.610469516074E−04	9.610469515368E−04
30	2	−14	12	1.741837353609E−03	1.741838878819E−03	1.741838878990E−03
30	2	−12	14	3.145381144114E−03	3.145384429887E−03	3.145384444067E−03
30	2	−10	16	5.713014458487E−03	5.713019249437E−03	5.713019508569E−03
30	2	−8	18	9.461086721272E−03	9.461146987153E−03	9.461143694266E−03
30	2	−6	20	2.825249919785E−02	2.825215025931E−02	2.825214335873E−02
30	2	−4	22	8.225223041839E−04	8.033676578715E−04	8.034028294691E−04
30	2	−2	24	−4.021604510865E−02	−4.067986999474E−02	−4.067958336874E−02
30	2	0	26	8.428871740799E−01	8.432289687375E−01	8.432306545436E−01
30	2	2	28	1.237497202556E+00	1.202531503930E+00	1.202530599134E+00
30	2	4	30	−7.293788189538E+00	−7.368422870064E+00	−7.368420896391E+00
30	2	6	32	−1.660458591396E+01	−1.596440454204E+01	−1.596440183564E+01
30	2	8	34	6.775794632054E−01	4.091453393401E+01	4.091453102328E+01
30	2	10	36	9.138903908292E+01	1.881260389012E+02	1.881260350363E+02
30	2	12	38	1.528259816376E+03	1.102086701962E+02	1.102086656476E+02
30	2	14	40	6.358866386921E+03	−8.444945803511E+02	−8.444945808309E+02
30	2	16	42	−9.457988143437E+02	−2.478127773453E+03	−2.478127776352E+03
30	2	18	44	4.031796088651E+03	−1.719989741133E+03	−1.719989739774E+03
30	2	20	46	3.729702671857E+05	7.099282454119E+03	7.099282455750E+03
l	p	q	k	McClain's first method Eq. (3.38) or Eq. (3.76)	McClain's second method Eq. (3.39) or Eq. (3.81)	Definite method Eq. (3.65)
30	2	−20	6	2.933228305424E−04	2.933228316203E−04	2.825986040143E−04
30	2	−18	8	5.309386709821E−04	5.309386721472E−04	5.177569571270E−04
30	2	−16	10	9.610469515536E−04	9.610469664404E−04	9.501852916158E−04
30	2	−14	12	1.741838879079E−03	1.741838950784E−03	1.728408552953E−03
30	2	−12	14	3.145384444781E−03	3.145384658439E−03	3.131415207725E−03
30	2	−10	16	5.713022055354E−03	5.713019138196E−03	5.698934401826E−03
30	2	−8	18	9.460862321659E−03	9.461143436325E−03	9.448181052947E−03
30	2	−6	20	2.823561066339E−02	2.825216233670E−02	2.824137621080E−02
30	2	−4	22	1.332378414479E−04	8.031214644615E−04	7.967125472328E−04
30	2	−2	24	−3.571464246330E−03	−4.067143402955E−02	−4.068994646337E−02
30	2	0	26	5.687372243357E−01	8.432070821565E−01	8.432180754044E−01
30	2	2	28	1.090101237776E+01	1.206212720880E+00	1.202518385562E+00
30	2	4	30	1.283979004667E+02	−7.365511398644E+00	−7.368430242571E+00

(continued)

Table 3.5 (continued)

l	p	q	k	McClain's first method Eq. (3.38) or Eq. (3.76)	McClain's second method Eq. (3.39) or Eq. (3.81)	Definite method Eq. (3.65)
30	2	6	32	−1.078978443250E+03	−1.504783954024E+01	−1.596441418996E+01
30	2	8	34	1.268956457029E+04	3.976137189384E+01	4.091452434365E+01
30	2	10	36	4.311563486128E+04	1.344069473763E+02	1.881260025841E+02
30	2	12	38	−1.462695561734E+05	−5.309894895613E+02	1.102086582301E+02
30	2	14	40	−9.518555070635E+05	−2.373497125213E+03	−8.444946464066E+02
30	2	16	42	−2.990934231992E+06	−2.828064882188E+03	−2.478127733697E+03
30	2	18	44	6.739078334877E+06	−1.189307052720E+04	−1.719989761921E+03
30	2	20	46	1.267869828572E+08	−1.100545306505E+04	7.099281584896E+03

3.4.2 Calculation Results for the Derivatives of Hansen Coefficients

The above several methods were used to calculate the derivatives of Hansen coefficients. The calculation results are given in Tables 3.6 and 3.7. It is seen that Wnuk's direct derivation method, Giacaglia's method and the definite integral method, i.e., Eqs. (3.88), (3.91) and (3.73), perform well in the calculation of the derivatives of Hansen coefficients, whereas Balmino's method, McClain's first method and McClain's second method, i.e., Eqs. (3.86), (3.79) and (3.84), are unstable for large eccentricities. In addition, although the calculation expressions for the derivatives of Hansen coefficients are different between Wnuk's method and Wnuk's direct derivation method, the corresponding calculation results of the derivatives of Hansen coefficients obtained with these two methods are in good agreement.

3.4.3 Computational Efficiencies

For the above several methods, the eccentricity functions $G_{lpq} = X_{l-2p+q}^{-(l+1),l-2p}$ ($l = 2 \sim 30, p = 0 \sim l, q = -2 \sim 2, e = 0.1$) and their derivatives were calculated. The computing times of the methods [5] are listed in Table 3.8 for reference. The basic configuration parameters of the computing device were an Intel® Xeon® W-2125 CPU @ 4.00 GHz (8 CPUs), basic frequency of approximately 4.0 GHz and 32,768 MB RAM.

Table 3.8 shows that the computational efficiencies of various methods vary greatly. Therefore, consideration should be given to the computational efficiency when applying these methods. In addition, it is noted that the summation ranges of the indexes in the various methods selected in this monograph may not be optimal. Numerical tests are required to choose appropriate summation ranges of the indexes

Table 3.6 Calculation example of the derivatives of Hansen coefficients ($e = 0.1$)

l	p	q	k	Derivation of Eq. (3.10), i.e., Eq. (3.86)	Derivation of Eq. (3.17), i.e., Eq. (3.88)	Derivation of Eq. (3.23), i.e., Eq. (3.91)
2	0	−6	−4	5.369329705503E−06	5.369329705503E−06	5.369329705503E−06
2	0	−5	−3	3.192054909411E−05	3.192054909411E−05	3.192054909411E−05
2	0	−4	−2	1.684350407088E−04	1.684350407088E−04	1.684350407088E−04
2	0	−3	−1	6.322334690673E−04	6.322334690673E−04	6.322334690673E−04
2	0	−2	0	0.000000000000E+00	0.000000000000E+00	0.000000000000E+00
2	0	−1	1	−4.981315652851E−01	−4.981315652851E−01	−4.981315652851E−01
2	0	0	2	−4.967572987486E−01	−4.967572987486E−01	−4.967572987486E−01
2	0	1	3	3.271279137771E+00	3.271279137771E+00	3.271279137771E+00
2	0	2	4	1.624081428131E+00	1.624081428131E+00	1.624081428131E+00
2	0	3	5	5.071858633876E−01	5.071858633876E−01	5.071858633876E−01
2	0	4	6	1.281295484913E−01	1.281295484913E−01	1.281295484912E−01
2	0	5	7	2.858242294190E−02	2.858242294190E−02	2.858242294190E−02
2	0	6	8	5.871524104337E−03	5.871524104337E−03	5.871524104337E−03
2	1	−6	−6	5.907429325512E−04	5.907429325512E−04	5.907429325512E−04
2	1	−5	−5	3.457539998965E−03	3.457539998965E−03	3.457539998965E−03
2	1	−4	−4	1.930064946432E−02	1.930064946432E−02	1.930064946432E−02
2	1	−3	−3	1.001597346921E−01	1.001597346921E−01	1.001597346921E−01
2	1	−2	−2	4.571341849402E−01	4.571341849402E−01	4.571341849402E−01
2	1	−1	−1	1.551661069825E+00	1.551661069825E+00	1.551661069825E+00
2	1	0	0	3.076332461767E−01	3.076332461767E−01	3.076332461767E−01
2	1	1	1	1.551661069825E+00	1.551661069825E+00	1.551661069825E+00
2	1	2	2	4.571341849402E−01	4.571341849402E−01	4.571341849402E−01
2	1	3	3	1.001597346921E−01	1.001597346921E−01	1.001597346921E−01
2	1	4	4	1.930064946432E−02	1.930064946432E−02	1.930064946432E−02
2	1	5	5	3.457539998965E−03	3.457539998965E−03	3.457539998965E−03
2	1	6	6	5.907429325512E−04	5.907429325512E−04	5.907429325512E−04
l	p	q	k	Derivation of Eq. (3.38), i.e., Eq. (3.79)	Derivation of Eq. (3.39), i.e., Eq. (3.84)	Definite integral method, i.e., Eq. (3.73)
2	0	−6	−4	5.369329705503E−06	5.369329705503E−06	5.369329705376E−06
2	0	−5	−3	3.192054909411E−05	3.192054909411E−05	3.192054909411E−05
2	0	−4	−2	1.684350407088E−04	1.684350407088E−04	1.684350407087E−04
2	0	−3	−1	6.322334690673E−04	6.322334690673E−04	6.322334690683E−04
2	0	−2	0	0.000000000000E+00	1.187588723943E−132	1.973976537784E−16
2	0	−1	1	−4.981315652851E−01	−4.981315652851E−01	−4.981315652851E−01
2	0	0	2	−4.967572987486E−01	−4.967572987486E−01	−4.967572987486E−01
2	0	1	3	3.271279137771E+00	3.271279137771E+00	3.271279137771E+00

(continued)

Table 3.6 (continued)

l	p	q	k	Derivation of Eq. (3.38), i.e., Eq. (3.79)	Derivation of Eq. (3.39), i.e., Eq. (3.84)	Definite integral method, i.e., Eq. (3.73)
2	0	2	4	1.624081428131E+00	1.624081428131E+00	1.624081428131E+00
2	0	3	5	5.071858633876E−01	5.071858633876E−01	5.071858633876E−01
2	0	4	6	1.281295484913E−01	1.281295484912E−01	1.281295484912E−01
2	0	5	7	2.858242294190E−02	2.858242294190E−02	2.858242294190E−02
2	0	6	8	5.871524104337E−03	5.871524104337E−03	5.871524104336E−03
2	1	−6	−6	5.907429325512E−04	5.907429325512E−04	5.907429325515E−04
2	1	−5	−5	3.457539998965E−03	3.457539998965E−03	3.457539998966E−03
2	1	−4	−4	1.930064946432E−02	1.930064946432E−02	1.930064946432E−02
2	1	−3	−3	1.001597346921E−01	1.001597346921E−01	1.001597346921E−01
2	1	−2	−2	4.571341849402E−01	4.571341849402E−01	4.571341849402E−01
2	1	−1	−1	1.551661069825E+00	1.551661069825E+00	1.551661069825E+00
2	1	0	0	3.076332461767E−01	3.076332461767E−01	3.076332461767E−01
2	1	1	1	1.551661069825E+00	1.551661069825E+00	1.551661069825E+00
2	1	2	2	4.571341849402E−01	4.571341849402E−01	4.571341849402E−01
2	1	3	3	1.001597346921E−01	1.001597346921E−01	1.001597346921E−01
2	1	4	4	1.930064946432E−02	1.930064946432E−02	1.930064946432E−02
2	1	5	5	3.457539998965E−03	3.457539998965E−03	3.457539998966E−03
2	1	6	6	5.907429325512E−04	5.907429325512E−04	5.907429325515E−04

in practical application. In fact, the selection of the summation ranges of the indexes used in various methods mainly depends on the eccentricity and the order of Hansen coefficients to be calculated.

3.5 Preliminary Analysis

The following preliminary conclusions are drawn from the above trial calculations of Hansen coefficients.

(1) For small orbital eccentricities ($e < 0.2$), various calculation methods satisfy the requirement of calculation accuracy.
(2) For large orbital eccentricities, Wnuk's [12] method (Wnuk's direct derivation method) and Giacaglia's [13] method provide relatively satisfactory calculation results.
(3) The main difficulty in the calculation of Hansen coefficients is the insufficient word length of the computer.
(4) A criterion is required to determine the stabilities of calculation methods.

Table 3.7 Calculation example of the derivatives of Hansen coefficients ($e = 0.6$)

l	p	q	k	Derivation of Eq. (3.10), i.e., Eq. (3.86)	Derivation of Eq. (3.17), i.e., Eq. (3.88)	Derivation of Eq. (3.23), i.e., Eq. (3.91)
30	2	−20	6	2.661736715817E−02	2.661739971325E−02	2.661739971322E−02
30	2	−18	8	4.676976221774E−02	4.676983070812E−02	4.676983070807E−02
30	2	−16	10	8.210444077516E−02	8.210458467688E−02	8.210458467478E−02
30	2	−14	12	1.440176512848E−01	1.440179529817E−01	1.440179529563E−01
30	2	−12	14	2.516108351830E−01	2.516114716841E−01	2.516114717679E−01
30	2	−10	16	4.486926962216E−01	4.486938663631E−01	4.486938738514E−01
30	2	−8	18	7.163356089092E−01	7.163418186674E−01	7.163416955212E−01
30	2	−6	20	1.492764035400E+00	1.492733619431E+00	1.492733131324E+00
30	2	−4	22	3.623905077806E+00	3.622459751844E+00	3.622441435702E+00
30	2	−2	24	−4.475814941376E+00	−4.513148158453E+00	−4.513130757659E+00
30	2	0	26	−2.539780445724E+00	−2.489694321637E+00	−2.489631742379E+00
30	2	2	28	1.480646532444E+02	1.455751879399E+02	1.455753422300E+02
30	2	4	30	2.028271543716E+02	1.896002464687E+02	1.896002830378E+02
30	2	6	32	−1.209845684990E+03	−1.143257190651E+03	−1.143257380209E+03
30	2	8	34	−6.515067615625E+03	−3.093514976359E+03	−3.093515320652E+03
30	2	10	36	−6.776044579618E+03	3.504528496240E+03	3.504529005598E+03
30	2	12	38	1.655520039208E+05	2.824022029641E+04	2.824022004950E+04
30	2	14	40	8.155317913596E+05	4.083095978474E+04	4.083095972905E+04
30	2	16	42	3.079822708888E+05	−6.047636033005E+04	−6.047636028013E+04
30	2	18	44	−1.380885271406E+05	−3.520368024379E+05	−3.520368020707E+05
30	2	20	46	3.932808553908E+07	−5.946661635728E+05	−5.946661626457E+05

l	p	q	k	Derivation of Eq. (3.38), i.e., Eq. (3.79)	Derivation of Eq. (3.39), i.e., Eq. (3.84)	Definite integral method, i.e., Eq. (3.73)
30	2	−20	6	2.661739971322E−02	2.661739986454E−02	2.551170563803E−02
30	2	−18	8	4.676983070807E−02	4.676983084617E−02	4.571388391172E−02
30	2	−16	10	8.210458467528E−02	8.210458642357E−02	8.121398238317E−02
30	2	−14	12	1.440179529535E−01	1.440179622038E−01	1.429985052762E−01
30	2	−12	14	2.516114716450E−01	2.516114984356E−01	2.509610672824E−01
30	2	−10	16	4.486941453135E−01	4.486938316979E−01	4.478407291042E−01
30	2	−8	18	7.163259095110E−01	7.163415852704E−01	7.154416090512E−01
30	2	−6	20	1.491503690212E+00	1.492735087742E+00	1.491900951589E+00
30	2	−4	22	3.500384424275E+00	3.622431759135E+00	3.621424715108E+00
30	2	−2	24	−2.387191291368E+00	−4.512588681010E+00	−4.514146455233E+00
30	2	0	26	−8.842378680474E+01	−2.496219237503E+00	−2.490747974808E+00
30	2	2	28	1.049968649974E+03	1.457966225199E+02	1.455743888852E+02
30	2	4	30	6.703722670208E+03	1.903142935908E+02	1.895991791796E+02

(continued)

Table 3.7 (continued)

l	p	q	k	Derivation of Eq. (3.38), i.e., Eq. (3.79)	Derivation of Eq. (3.39), i.e., Eq. (3.84)	Definite integral method, i.e., Eq. (3.73)
30	2	6	32	−1.142780533955E+05	−1.080511472251E+03	−1.143258511742E+03
30	2	8	34	1.524066036680E+06	−3.305353234368E+03	−3.093515957706E+03
30	2	10	36	4.499116834328E+06	−1.781507310351E+03	3.504527699258E+03
30	2	12	38	−3.172345379743E+07	−3.022726111072E+04	2.824021906043E+04
30	2	14	40	−7.087515494078E+07	−1.313244256416E+05	4.083096159104E+04
30	2	16	42	−4.736466724733E+08	3.720257340098E+04	−6.047636001794E+04
30	2	18	44	−5.366466556619E+08	−1.614577142640E+06	−3.520368051023E+05
30	2	20	46	−8.724955200509E+09	−7.396491473777E+06	−5.946660904052E+05

Table 3.8 Computing times of various methods

Method	Expression of Hansen coefficients	Expression of the Derivatives	Time (s)
Balmino's method [4]	(3.10)	(3.86)	364.127
Wnuk's method [12]	(3.17) or (3.87)	(3.74b)	49.317
Wnuk's direct derivation method [5]	(3.17) or (3.87)	(3.88)	16.715
Giacaglia's method [13]	(3.23) or (3.90)	(3.91)	4.827
McClain's first method [1, 16] (double precision computation)	(3.38) or (3.76)	(3.79)	0.031
McClain's first method [1, 16] (quadruple precision computation)	(3.38) or (3.76)	(3.79)	0.329
McClain's second method [1, 16] (double precision computation)	(3.39) or (3.81)	(3.84)	0.422
McClain's second method [1, 16] (quadruple precision computation)	(3.39) or (3.81)	(3.84)	11.607
Definite integral method	(3.65)	(3.73)	0.719

All the instabilities of the calculations of Hansen coefficients occur in the case of large orbital eccentricities, whereas there is no such problem for small orbital eccentricities. For the calculations of Hansen coefficients with large orbital eccentricities, the main reason for the loss of significant digits is as follows. The Hansen coefficients are calculated through the summation of series; i.e., $y = \sum x_i$. There are factorials and binomial coefficients in x_i, whose magnitudes vary so greatly for large orbital eccentricities that calculation accuracy is lost in the summation. The greatest issue is that the positive and negative cancel for the numbers with a large order of magnitude in the summing operation, and the rest is a decimal (it is the calculation result of Hansen coefficient). If this decimal is smaller than the last digit value of the maximum number, there is no significant digit in this case, and the instability problem appears. For the direct methods of calculating Hansen coefficients, a summation of series is inevitable. There is thus a need for a method with small differences of magnitudes of x_i. Of course, it is an effective method to adopt quadruple precision computation.

In the above, the stabilities of the direct methods of calculating Hansen coefficients were determined by conducting example tests and following the rule that the minority follows the majority. Obviously, this is not sufficient, because a method that is stable in one example is not necessarily stable in other examples. A criterion is thus required to determine the stabilities of calculation methods. In Chap. 4, we further investigate this problem.

Perturbation calculations without singularities are required considering that small orbital eccentricities are adopted for most satellites. The kernels of Hansen coefficients $K_t^{n,s} = e^{-|t-s|} X_t^{n,s}$ are needed for perturbation calculations without singularities, and they are derived by extracting the factor $e^{|t-s|}$ from Hansen coefficients $X_t^{n,s}$. Moreover, there is no small denominator problem with the orbital eccentricity e being zero in the kernels of Hansen coefficients $K_t^{n,s}$ themselves. Wnuk's [12] method (Wnuk's direct derivation method), i.e., Eq. (3.17) or Eq. (3.87), does not meet this requirement. Hence, Wnuk's method cannot be used for the calculation of the kernels of Hansen coefficients and their derivatives and is unavailable for perturbation calculations without singularities. Although the calculation results of McClain's first method [1, 16] with double precision computation are poor for higher-order Hansen coefficients of large eccentricities, the method can be used for the convenient calculation of the kernels of Hansen coefficients $K_t^{n,s}$ and their derivatives $\frac{dK_t^{n,s}}{de}$ at small eccentricities and is available for perturbation calculations without singularities. McClain's first method [1, 16] with quadruple precision computation covers the calculations of Hansen coefficients for both large and small orbital eccentricities and is thus a recommended choice.

In addition, the calculation of Hansen coefficients is a part of perturbation calculations. Therefore, the calculation of Hansen coefficients should be studied together with perturbation calculations. We might ask, in perturbation calculations with double precision computation, can two methods of calculating Hansen coefficients be used? For example, in the case of double-precision computation, at large eccentricities, on the basis of Keplerian elements, Wnuk's [12] method (Wnuk's direct derivation method) might be used, whereas at small eccentricities, on the basis of non-singular elements, McClain's first method [1, 16] might be used. This might be a reasonable choice, before a better method is developed. Of course, it is not convenient to use two element systems in one orbit calculation software application.

References

1. McClain WD (1978) A recursively formulated first-order semianalytic artificial satellite theory based on the generalized method of averaging. Volume II. The explicit development of the first-order averaged equations of motion for the nonspherical gravitational and nonresonant third-body perturbations. NASA CR-156783

2. Izsak IG, Gerard JM, Efimba R, Barnett MP (1964) Construction of newcomb operators on a digital computer. SAO special report No. 140. Smithsonian Astrophysical Observatory
3. Cherniack JR (1972) Computation of Hansen coefficients. SAO special report No. 346. Smithsonian Astrophysical Observatory
4. Balmino G (2005) Geodetic satellite orbits in the earth's gravity field. Tutorials on theoretical or methodological aspects in Gravimetry. International Gravimetric Bureau
5. Wu LD, Zhang MJ (2022) Direct calculation methods of Hansen coefficients and their derivatives. Chin Astron Astrophy 46(1):137–151
6. Gaposchkin EM (1973) 1973 Smithsonian standard earth (III). SAO special report No. 353. Smithsonian Astrophysical Observatory, Cambridge
7. Plummer HC (1918) An introductory treatise on dynamical astronomy. Cambridge University Press, London
8. Branham RL (1990) Recursive calculation of Hansen coefficients. Celest Mech Dyn Astron 49(2):209–217
9. Plummer HC (1960) An introductory treatise on dynamical astronomy. Dover, New York
10. Breiter S, Métris G, Vokrouhlický D (2004) Generalized Hansen coefficients. Celest Mech Dyn Astron 88(2):153–161
11. Aksenov EP (1986) Special functions in celestial mechanics. Nauka, Moscow
12. Wnuk E (1997) Highly eccentric satellite orbits. Adv Space Res 19(11):1735–1740
13. Giacaglia GEO (1976) A note on Hansen's coefficients in satellite theory. Celest Mech 14(4):515–523
14. Brouwer D, Clemence GM (1961) Methods of celestial mechanics. Academic Press, New York and London
15. Harrison J (2009) Fast and accurate Bessel function computation. In: Proceedings of the 19th IEEE international symposium on computer arithmetic, pp 104–113
16. Proulx RJ, McClain WD (1988) Series representations and rational approximations for Hansen coefficients. J Guid 11(4):313–319
17. Oesterwinter D (1989) Semianalytic satellite theory. NSWC TR 89-109
18. Kaula WM (1966) Theory of satellite geodesy: applications of satellites to geodesy. Blaisdell Publishing Company
19. Tisserand F (1889) Traité de Mécanique Céleste. Gauthier-Villars et fils, Paris
20. Laskar J (2005) Note on the generalized Hansen and Laplace coefficients. Celest Mech Dyn Astron 91(3–4):351–356
21. Laskar J, Robutel P (1995) Stability of the planetary three-body problem. Celest Mech Dyn Astron 62(3):193–217
22. Lion G, Métris G (2013) Two algorithms to compute Hansen-like coefficients with respect to the eccentric anomaly. Adv Space Res 51(1):1–9
23. Challe A, Laclaverie JJ (1969) Disturbing function and analytical solution of the problem of the motion of a satellite. Astron Astrophys 3(3):15–28

Chapter 4
Recursion of Hansen Coefficients

This chapter investigates recursion calculation methods of Hansen coefficients $X_k^{n,m}$ and their derivatives for an arbitrary index k, which is another important research issue addressed in this monograph.

4.1 Recursion Relations of Hansen Coefficients

Giacaglia and McClain presented many relations for the recursion calculation of Hansen coefficients $X_k^{n,m}$ [1, 2]. In this section, the recursion formulae of Hansen coefficients given by Giacaglia and McClain are reviewed, the fact that there are only five independent recursion formulae among those given by Giacaglia and McClain is pointed out, and a remarkable characteristic of the recursion formulae of Hansen coefficients is analyzed briefly.

4.1.1 Giacaglia's Recursion Formulae [1]

The definition of Hansen coefficient $X_k^{n,m}$ is

$$\left(\frac{r}{a}\right)^n \exp(jmf) = \sum_{k=-\infty}^{\infty} X_k^{n,m} \exp(jkM) \tag{4.1}$$

where r is the geocentric distance, a is the semi-major axis, f is the true anomaly, M is the mean anomaly, and $j = \sqrt{-1}$. By directly differentiating Eq. (4.1) with respect to the mean anomaly M, we obtain

L.D Wu and M.J Zhang, *Hansen Coefficients in Satellite Orbital Dynamics*,
Springer Aerospace Technology, https://doi.org/10.1007/978-981-97-0456-9_4

$$\begin{aligned}
&\frac{-ne}{2\sqrt{1-e^2}}\left(\frac{r}{a}\right)^{n-1}\left[\exp(j(m+1)f)-\exp(j(m-1)f)\right] \\
&\quad + m\sqrt{1-e^2}\left(\frac{r}{a}\right)^{n-2}\exp(jmf) \\
&= \sum_{k=-\infty}^{\infty} kX_k^{n,m}\exp(jkM)
\end{aligned} \tag{4.2}$$

Applying definition Eq. (4.1) to Eq. (4.2), a recursion formula of Hansen coefficients $X_k^{n,m}$ is obtained immediately as

$$kX_k^{n,m} = m\sqrt{1-e^2}X_k^{n-2,m} + \frac{ne}{2\sqrt{1-e^2}}\left(X_k^{n-1,m-1} - X_k^{n-1,m+1}\right) \tag{4.3}$$

Equation (4.3) holds for all n, m and k. Replacing the index n by $n+2$, i.e., $n \Rightarrow n+2$, yields

$$m\left(1-e^2\right)X_k^{n,m} = k\sqrt{1-e^2}X_k^{n+2,m} - \frac{(n+2)e}{2}\left(X_k^{n+1,m-1} - X_k^{n+1,m+1}\right)$$

Transforming Hansen coefficients into eccentricity functions, the recursion formula of eccentricity functions is obtained as

$$\begin{aligned}
(l-2p)\left(1-e^2\right)G_{lpq} &= (l-2p+q)\sqrt{1-e^2}G_{l-2,p-1,q} \\
&\quad - \frac{(l-1)e}{2}\left(G_{l-1,p-1,q-1} - G_{l-1,p,q+1}\right)
\end{aligned} \tag{4.4}$$

We rewrite Eq. (4.1) as

$$\begin{aligned}
\left(\frac{r}{a}\right)^{n+1}\frac{a}{r}\exp(jmf) &= \frac{1}{1-e^2}\left(\frac{r}{a}\right)^{n+1}\exp(jmf) \\
&\quad + \frac{e}{2\left(1-e^2\right)}\left(\frac{r}{a}\right)^{n+1}\left[\exp(j(m+1)f)+\exp(j(m-1)f)\right] \\
&= \sum_{k=-\infty}^{\infty} X_k^{n,m}\exp(jkM)
\end{aligned}$$

and then by applying definition Eq. (4.1) to the above formula, a recursion formula of Hansen coefficients $X_k^{n,m}$ is obtained immediately as

$$\left(1-e^2\right)X_k^{n,m} = X_k^{n+1,m} + \frac{e}{2}\left(X_k^{n+1,m+1} + X_k^{n+1,m-1}\right) \tag{4.5}$$

Further rewriting the factor $\left(\frac{r}{a}\right)^n$ in Eq. (4.1) as $\left(\frac{r}{a}\right)^n = \left(\frac{r}{a}\right)^{n+2}\left(\frac{a}{r}\right)^2$, a recursion formula of Hansen coefficients $X_k^{n,m}$ is similarly obtained as

$$
\begin{aligned}
\left(1-e^2\right)^2 X_k^{n,m} = \left(1+\frac{e^2}{2}\right) X_k^{n+2,m} + e\left(X_k^{n+2,m+1} + X_k^{n+2,m-1}\right) \\
+ \frac{e^2}{4}\left(X_k^{n+2,m+2} + X_k^{n+2,m-2}\right)
\end{aligned}
\tag{4.6}
$$

Replacing the index n in Eq. (4.5) by $n+1$, i.e., $n \Rightarrow n+1$, yields

$$
e\left(X_k^{n+2,m+1} + X_k^{n+2,m-1}\right) = 2\left(1-e^2\right) X_k^{n+1,m} - 2X_k^{n+2,m}
$$

Substituting Eq. (4.5) into the above equation to eliminate $X_k^{n+1,m}$ yields

$$
\begin{aligned}
e\left(X_k^{n+2,m+1} + X_k^{n+2,m-1}\right) = 2\left(1-e^2\right)^2 X_k^{n,m} - \left(1-e^2\right)e\left(X_k^{n+1,m+1} + X_k^{n+1,m-1}\right) \\
- 2X_k^{n+2,m}
\end{aligned}
$$

Then, by substituting the above equation into Eq. (4.6) to eliminate $e\left(X_k^{n+2,m+1} + X_k^{n+2,m-1}\right)$, a recursion formula of Hansen coefficients $X_k^{n,m}$ is obtained immediately as

$$
\begin{aligned}
\left(1-e^2\right)^2 X_k^{n,m} = \left(1-\frac{e^2}{2}\right) X_k^{n+2,m} \\
+ \left(1-e^2\right)e\left(X_k^{n+1,m+1} + X_k^{n+1,m-1}\right) - \frac{e^2}{4}\left(X_k^{n+2,m+2} + X_k^{n+2,m-2}\right)
\end{aligned}
\tag{4.7}
$$

Transforming Hansen coefficients into eccentricity functions $G_{lpq} = X_{l-2p+q}^{-(l+1),l-2p}$, the recursion formula of eccentricity functions is obtained as

$$
\begin{aligned}
\left(1-e^2\right)^2 G_{lpq} = \left(1-\frac{e^2}{2}\right) G_{l-2,p-1,q} + \left(1-e^2\right)e\left(G_{l-1,p-1,q-1} + G_{l-1,p,q+1}\right) \\
- \frac{e^2}{4}\left(G_{l-2,p-2,q-2} + G_{l-2,p,q+2}\right)
\end{aligned}
\tag{4.8}
$$

Equations (4.3), (4.5) and (4.7) are the important recursion formulae of Hansen coefficients $X_k^{n,m}$.

4.1.2 *McClain's Recursion Formulae [2]*

McClain's recursion formulae are derived using the classic Hansen's method. Let

$$\rho = \frac{r}{a}, \quad x = \exp(jf), \quad z = \exp(jM)$$

Then

$$z\frac{d}{dz}\rho^n x^s = n\rho^{n-1}x^s z\frac{d\rho}{dz} + s\rho^n x^{s-1} z\frac{dx}{dz} \tag{4.9}$$

Using the relations

$$z\frac{d}{dz} = z\frac{dx}{dz}\frac{d}{dx} \tag{4.10}$$

$$\frac{dx}{dz} = \frac{x}{z}\frac{a^2}{r^2}\cos\phi \tag{4.11}$$

$$\begin{aligned} x\frac{d\rho}{dx} &= \frac{1}{2}\frac{r^2}{a^2}\frac{\sin\phi}{\cos^2\phi}\left(\frac{1}{x} - x\right) \\ z\frac{d\rho}{dz} &= z\frac{d\rho}{dx}\frac{dx}{dz} = \frac{a^2}{r^2}\cos\phi x\frac{d\rho}{dx} = \frac{1}{2}\frac{\sin\phi}{\cos\phi}\left(\frac{1}{x} - x\right) \end{aligned} \tag{4.12}$$

Eq. (4.9) is rewritten as

$$z\frac{d}{dz}\rho^n x^s = \frac{n\sin\phi}{2\cos\phi}\rho^{n-1}\left(x^{s-1} - x^{s+1}\right) + s\cos\phi\rho^{n-2}x^s \tag{4.13}$$

In Eqs. (4.11)–(4.13), $\sin\phi = e$ and $\cos\phi = \sqrt{1-e^2}$.
According to the orbital equation,

$$\rho - \frac{\cos^2\phi}{1 + \frac{\sin\phi}{2}\left(x + x^{-1}\right)} \equiv 0 \tag{4.14}$$

i.e.,

$$\sin\phi\left(x + x^{-1}\right) - 2\cos^2\phi\rho^{-1} + 2 = 0 \tag{4.15}$$

Multiplying Eq. (4.15) by $\pm\frac{n\rho^{n-1}x^s}{2\cos\phi}$ and $\frac{s\rho^{n-1}x^s}{2\cos\phi}$ yields

$$\begin{aligned} &\frac{n\rho^{n-1}x^s}{2\cos\phi}\sin\phi\left(x + x^{-1}\right) - n\rho^{n-2}x^s\cos\phi + \frac{n\rho^{n-1}x^s}{\cos\phi} = 0 \\ -&\frac{n\rho^{n-1}x^s}{2\cos\phi}\sin\phi\left(x + x^{-1}\right) + n\rho^{n-2}x^s\cos\phi - \frac{n\rho^{n-1}x^s}{\cos\phi} = 0 \\ &\frac{s\rho^{n-1}x^s}{2\cos\phi}\sin\phi\left(x + x^{-1}\right) - s\rho^{n-2}x^s\cos\phi + \frac{s\rho^{n-1}x^s}{\cos\phi} = 0 \end{aligned}$$

As the left sides of the above three formulae are equal to zero, they can be added to the right side of Eq. (4.13), and we have

$$z\frac{d}{dz}\rho^n x^s = \frac{n\sin\phi}{\cos\phi}\rho^{n-1}x^{s-1} + (s-n)\cos\phi\rho^{n-2}x^s + \frac{n}{\cos\phi}\rho^{n-1}x^s \tag{4.16}$$

$$z\frac{d}{dz}\rho^n x^s = -\frac{n\sin\phi}{\cos\phi}\rho^{n-1}x^{s+1} + (s+n)\cos\phi\rho^{n-2}x^s - \frac{n}{\cos\phi}\rho^{n-1}x^s \tag{4.17}$$

$$z\frac{d}{dz}\rho^n x^s = \frac{(n+s)\sin\phi}{2\cos\phi}\rho^{n-1}x^{s-1} - \frac{(n-s)\sin\phi}{2\cos\phi}\rho^{n-1}x^{s+1} + \frac{s}{\cos\phi}\rho^{n-1}x^s \tag{4.18}$$

It follows from Eq. (4.14) that

$$\frac{\sin\phi}{2\cos^2\phi}\rho\left(x+x^{-1}\right) + \frac{\rho}{\cos^2\phi} = 1 \tag{4.19}$$

As the left side of Eq. (4.19) is equal to 1, it can be multiplied with the right side of Eq. (4.18) to get

$$\begin{aligned} z\frac{d}{dz}\rho^n x^s &= \frac{(n+s)\sin^2\phi}{4\cos^3\phi}\rho^n x^{s-2} + \frac{(n+2s)\sin\phi}{2\cos^3\phi}\rho^n x^{s-1} \\ &+ \frac{2s+s\sin^2\phi}{2\cos^3\phi}\rho^n x^s - \frac{(n-2s)\sin\phi}{2\cos^3\phi}\rho^n x^{s+1} - \frac{(n-s)\sin^2\phi}{4\cos^3\phi}\rho^n x^{s+2} \end{aligned} \tag{4.20}$$

Using the operator $z\frac{d}{dz}$ to perform differential calculation of Eqs. (4.16) and (4.17), Hansen derived that

$$\begin{aligned} z^2\frac{d^2}{dz^2}\rho^n x^s + z\frac{d}{dz}\rho^n x^s &= \left[n(n-2)+s^2\right]\cos^2\phi\rho^{n-4}x^s \\ &- n(2n-3)\rho^{n-3}x^s + n(n-1)\rho^{n-2}x^s \\ &+ s(n-1)\sin\phi\rho^{n-3}\left(x^{s-1}-x^{s+1}\right) \end{aligned} \tag{4.21}$$

We multiply the left side of Eq. (4.15) by $s(n-1)\rho^{n-3}x^s$, and the result is then added to and subtracted from the right side of Eq. (4.21) to obtain

$$\begin{aligned} z^2\frac{d^2}{dz^2}\rho^n x^s + z\frac{d}{dz}\rho^n x^s &= \left[n(n-2)+s^2-2s(n-1)\right]\cos^2\phi\rho^{n-4}x^s \\ &- [n(2n-3)-2s(n-1)]\rho^{n-3}x^s \\ &+ n(n-1)\rho^{n-2}x^s + 2s(n-1)\sin\phi\rho^{n-3}x^{s-1} \end{aligned} \tag{4.22}$$

$$z^2\frac{d^2}{dz^2}\rho^n x^s + z\frac{d}{dz}\rho^n x^s = \left[n(n-2)+s^2+2s(n-1)\right]\cos^2\phi\rho^{n-4}x^s$$
$$-\left[n(2n-3)+2s(n-1)\right]\rho^{n-3}x^s$$
$$+n(n-1)\rho^{n-2}x^s - 2s(n-1)\sin\phi\rho^{n-3}x^{s+1} \quad (4.23)$$

We substitute the series expression of $\rho^n x^s$ and its first-order and second-order derivatives

$$\rho^{\pm n}x^s = \sum_t X_t^{\pm n,s} z^t \quad (4.24a)$$

$$z\frac{d}{dz}\rho^{\pm n}x^s = \sum_t tX_t^{\pm n,s} z^t \quad (4.24b)$$

$$z\frac{d^2}{dz^2}\rho^{\pm n}x^s = \sum_t t(t-1)X_t^{\pm n,s} z^t \quad (4.24c)$$

into the differential Eqs. (4.13), (4.16)–(4.18) and (4.20)–(4.23) and perform a unified disposal of parameters and indexes according to

$$\sin\phi = e, \quad \cos\phi = \sqrt{1-e^2}, \quad s \Rightarrow m, \quad t \Rightarrow k$$

The recursion relations of Hansen coefficients $X_k^{n,m}$ are then obtained immediately as

$$neX_k^{n-1,m-1} - neX_k^{n-1,m+1} + 2m\left(1-e^2\right)X_k^{n-2,m} - 2k\sqrt{1-e^2}X_k^{n,m} = 0 \quad (4.25)$$

$$neX_k^{n-1,m-1} + (m-n)\left(1-e^2\right)X_k^{n-2,m} + nX_k^{n-1,m} - k\sqrt{1-e^2}X_k^{n,m} = 0 \quad (4.26)$$

$$-neX_k^{n-1,m+1} + (m+n)\left(1-e^2\right)X_k^{n-2,m} - nX_k^{n-1,m} - k\sqrt{1-e^2}X_k^{n,m} = 0 \quad (4.27)$$

$$(n+m)eX_k^{n-1,m-1} - (n-m)eX_k^{n-1,m+1} + 2mX_k^{n-1,m} - 2k\sqrt{1-e^2}X_k^{n,m} = 0 \quad (4.28)$$

$$(n+m)e^2X_k^{n,m-2} + 2(n+2m)eX_k^{n,m-1} + \left[4m+2me^2-4k\left(1-e^2\right)^{3/2}\right]X_k^{n,m}$$
$$-2(n-2m)eX_k^{n,m+1} - (n-m)e^2X_k^{n,m+2} = 0 \quad (4.29)$$

$$\left[n(n-2)+m^2\right]\left(1-e^2\right)X_k^{n-4,m} - n(2n-3)X_k^{n-3,m} + n(n-1)X_k^{n-2,m}$$
$$+m(n-1)eX_k^{n-3,m-1} - m(n-1)eX_k^{n-3,m+1} - k^2X_k^{n,m} = 0 \quad (4.30)$$

$$\left[n(n-2)+m^2-2m(n-1)\right]\left(1-e^2\right)X_k^{n-4,m} - \left[n(2n-3)-2m(n-1)\right]X_k^{n-3,m}$$

$$+n(n-1)X_k^{n-2,m} + 2m(n-1)eX_k^{n-3,m-1} - k^2X_k^{n,m} = 0 \tag{4.31}$$

$$\begin{aligned}&\big[n(n-2)+m^2+2m(n-1)\big]\big(1-e^2\big)X_k^{n-4,m} - [n(2n-3)+2m(n-1)]X_k^{n-3,m}\\ &+n(n-1)X_k^{n-2,m} - 2m(n-1)eX_k^{n-3,m+1} - k^2X_k^{n,m} = 0\end{aligned} \tag{4.32}$$

According to recursion relations Eqs. (4.25) and (4.28), we have

$$eX_k^{n-1,m-1} + eX_k^{n-1,m+1} - 2\big(1-e^2\big)X_k^{n-2,m} + 2X_k^{n-1,m} = 0 \tag{4.33}$$

If the index n in Eq. (4.33) is replaced by $n+2$, i.e., $n \Rightarrow n+2$, then Eq. (4.33) is Giacaglia's recursion formula Eq. (4.5).

Replacing the index n in Eq. (4.25) by $n-2$, i.e., $n \Rightarrow n-2$, we have

$$\Big(eX_k^{n-3,m-1} - eX_k^{n-3,m+1}\Big) = -\frac{2m\big(1-e^2\big)}{n-2}X_k^{n-4,m} + \frac{2k\sqrt{1-e^2}}{n-2}X_k^{n-2,m}$$

Substituting the above formula into Eq. (4.30) to eliminate the term $\Big[m(n-1)eX_k^{n-3,m-1} - m(n-1)eX_k^{n-3,m+1}\Big]$ gives the recursion formula

$$\begin{aligned}&n\big[(n-2)^2-m^2\big]\big(1-e^2\big)X_k^{n-4,m} - n(n-2)(2n-3)X_k^{n-3,m}\\ &+(n-1)\Big[n(n-2)+2km\sqrt{1-e^2}\Big]X_k^{n-2,m} - (n-2)k^2X_k^{n,m} = 0\end{aligned} \tag{4.34}$$

The superscript m in Eq. (4.34) is the same and there is no denominator of e. Equation (4.34) is especially suited to perturbation theory without singularities, but it cannot be transformed into a recursion formula of eccentricity functions.

Other recursion formulae can be obtained by combining the above recursion formulae. These recursion formulae hold for all n, m and k. It is not difficult to find that Eq. (4.25) is equivalent to Eq. (4.3); the result obtained by the subtraction of Eqs. (4.25) and (4.28) is Eq. (4.33) and equivalent to Eq. (4.5); the result obtained by the addition of Eqs. (4.26) and (4.27) is equivalent to Eq. (4.3); the result obtained by the subtraction of Eqs. (4.26) and (4.27) is equivalent to Eq. (4.5); the result obtained by the addition of Eqs. (4.31) and (4.32) is Eq. (4.30); the result obtained by the subtraction of Eqs. (4.31) and (4.32) is equivalent to Eq. (4.5); and Eq. (4.34) can be derived from Eqs. (4.25) and (4.30). Therefore, there are only three independent recursion formulae among Eqs. (4.3), (4.5), (4.25)–(4.28) and (4.30)–(4.34). In McClain's recursion formulae, there is no formula equivalent to Eq. (4.7). Ultimately, there are only five independent recursion formulae among those given by Giacaglia and McClain, namely Eqs. (4.3), (4.5), (4.7), (4.29) and (4.34).

4.1.3 A Characteristic of the Recursion Formulae

An examination of the above recursion formulae of Hansen coefficients (including the recursion formulae presented in other references, such as in Ref. [3]) reveals that the formulae possess a remarkable characteristic in that the subscripts k of Hansen coefficients $X_k^{n,m}$ in the recursion formulae are the same. Of course, in this way, the recursion formulae have the advantage that the recursive process is relatively simple. When $k = 0$, the recursion formulae in Chap. 2 can be used in the calculation. However, because we need to calculate Hansen coefficients of many k values, the recursion of Hansen coefficients for each k value must be performed. Each time the k value changes, the recursion is restarted, and the corresponding initial values of the recursion are recalculated, which severely reduces the computational efficiency of the recursion.

In view of this remarkable characteristic of the recursion formulae of Hansen coefficients, except for the batch recursion of eccentricity functions, the recursion of Hansen coefficients investigated in this chapter is only for Hansen coefficients with the same k values and not all Hansen coefficients.

In addition, Hansen coefficients derived using the recursion methods with the index k equal to a constant are not necessarily what we need. As the Hansen coefficients $X_k^{-(n+1),m}$ and $X_k^{n,m}$ contain the $e^{|k-m|}$ factor, their magnitudes are not the same for the same index k but the different index m. For example, there is a difference of e^{2n} between $X_k^{-(n+1),n}$ and $X_k^{-(n+1),-n}$ when $k \geq n$. If $n = 10$ and $e = 0.1$, there is a difference of 20 orders of magnitude between these two Hansen coefficients. However, the two Hansen coefficients $X_k^{-(n+1),n}$ and $X_k^{-(n+1),-n}$ are calculated simultaneously when adopting the recursion methods of Giacaglia [1] and McClain [2], which is obviously unnecessary and unreasonable.

4.2 Recursion Calculation of Hansen Coefficients

In satellite dynamics, the order of Hansen coefficients involved in luni-solar perturbations is generally low, and recursion is thus not adopted. In the following, we mainly investigate the recursion of Hansen coefficients involved in the Earth's gravitational field. Here, the general term of Hansen coefficients is $X_k^{-(n+1),m}$, where n is a positive integer, $m = n - 2p$, the value range of n is $2 \sim N$, the value range of p is $0 \sim n$, and the value range of m is $-n \sim n$. The recursion of Hansen coefficients $X_k^{-(n+1),m}$ is expected to be generally performed in the direction of n.

4.2.1 Basic Recursion Formulae

As stated above, only five of the recursion formulae of Hansen coefficients given by Giacaglia and McClain [1, 2] (presented in Sect. 4.1), namely Eqs. (4.3), (4.5), (4.7), (4.29) and (4.34), are independent. As recursion formula Eq. (4.34) involves four Hansen coefficients with different indexes n, it requires more initial values, and it is inconvenient to perform recursion. Equations (4.3), (4.5), (4.7) and (4.29) are thus chosen as four basic recursion formulae and further expressed in the form of the general term $X_k^{-(n+1),m}$ [4]:

$$m\sqrt{1-e^2}X_k^{-(n+2),m} = \frac{ne}{2\sqrt{1-e^2}}\left[X_k^{-(n+1),m-1} - X_k^{-(n+1),m+1}\right] + kX_k^{-n,m} \quad \text{(R1)}$$

$$\left(1-e^2\right)X_k^{-(n+1),m} = X_k^{-n,m} + \frac{e}{2}\left(X_k^{-n,m+1} + X_k^{-n,m-1}\right) \quad \text{(R2)}$$

$$\begin{aligned}\left(1-e^2\right)^2X_k^{-(n+2),m} &= \left(1-e^2\right)e\left[X_k^{-(n+1),m+1} + X_k^{-(n+1),m-1}\right] \\ &\quad + \left(1-\frac{e^2}{2}\right)X_k^{-n,m} - \frac{e^2}{4}\left(X_k^{-n,m+2} + X_k^{-n,m-2}\right) \quad \text{(R3)}\end{aligned}$$

$$\begin{aligned}&(m-n-1)e^2X_k^{-(n+1),m-2} + 2(2m-n-1)eX_k^{-(n+1),m-1} \\ &+ \left[4m + 2me^2 - 4k\left(1-e^2\right)^{3/2}\right]X_k^{-(n+1),m} \\ &+ 2(n+2m+1)eX_k^{-(n+1),m+1} + (n+m+1)e^2X_k^{-(n+1),m+2} = 0 \quad \text{(R4)}\end{aligned}$$

4.2.2 Classification of Recursions

According to the requirements of recursion, there are many types of classifications for the recursion of Hansen coefficients. The common classifications are summarized in the following five classes.

(1) **Recursions in the directions of n and m**

Basic recursion formulae Eqs. (R1), (R2) and (R3) can be used to perform the recursion in the direction of n, whereas basic recursion formula Eq. (R4) can be used to perform the recursion in the direction of m.

(2) **Forward and backward recursions**

The recursions implemented using the basic recursion formulae Eqs. (R1), (R2), (R3) and (R4) are illustrated below. The recursion of Eq. (R1) forms a rhombus and can be in either forward or backward direction.

$$\begin{array}{ccc} & X_k^{-n,m} & \\ X_k^{-(n+1),m-1} & & X_k^{-(n+1),m+1} \\ & X_k^{-(n+2),m} & \end{array} \quad \begin{array}{l} X_k^{-n,m}, X_k^{-(n+1),m-1}, X_k^{-(n+1),m+1} \Rightarrow X_k^{-(n+2),m} \\ X_k^{-n,m} \Leftarrow X_k^{-(n+1),m-1}, X_k^{-(n+1),m+1}, X_k^{-(n+2),m} \end{array}$$

The recursion of Eq. (R2) has the pattern that one Hansen coefficient is derived from three other Hansen coefficients and is in the forward direction.

$$\begin{array}{ccc} X_k^{-n,m-1} & X_k^{-n,m} & X_k^{-n,m+1} \\ & X_k^{-(n+1),m} & \end{array} \quad X_k^{-n,m-1}, X_k^{-n,m}, X_k^{-n,m+1} \Rightarrow X_k^{-(n+1),m}$$

The recursion of Eq. (R3) forms an inverted tower and is in the forward direction.

$$\begin{array}{ccccc} X_k^{-n,m-2} & & X_k^{-n,m} & & X_k^{-n,m+2} \\ & X_k^{-(n+1),m-1} & & X_k^{-(n+1),m+1} & \\ & & X_k^{-(n+2),m} & & \end{array}$$

$$X_k^{-n,m-2}, X_k^{-n,m}, X_k^{-n,m+2}, X_k^{-(n+1),m-1}, X_k^{-(n+1),m+1} \Rightarrow X_k^{-(n+2),m}$$

The recursion of Eq. (R4) has a linear form and is in either the forward or backward direction.

$$X_k^{-n,m-2}, X_k^{-n,m-1}, X_k^{-n,m}, X_k^{-n,m+1} \Rightarrow X_k^{-n,m+2}$$
$$X_k^{-n,m-2} \Leftarrow X_k^{-n,m-1}, X_k^{-n,m}, X_k^{-n,m+1}, X_k^{-n,m+2}$$

It is thus seen that the recursions implemented using the basic recursion formulae Eqs. (R1), (R2) and (R3) are applicable for forward recursion in the direction of n, and only the basic recursion formula Eq. (R1) can be used for backward recursion in the direction of n.

(3) **Application for recursion of eccentricity functions**

The basic recursion formulae Eqs. (R1) and (R3) are applicable for the recursion of eccentricity functions, whereas the basic recursion formulae Eqs. (R2) and (R4) are not applicable.

(4) **One line of initial values and two lines of initial values**

One line of initial values is required to perform recursion using the basic recursion formula Eq. (R2), whereas two lines of initial values are required for the basic recursion formulae Eqs. (R1) and (R3).

(5) **Auxiliary recursion required in forward recursion in the direction of** n

We consider the following example of the recursion process.

$$\begin{array}{c} X_k^{-3,-2}, X_k^{-3,-1}, X_k^{-3,0}, X_k^{-3,1}, X_k^{-3,2} \\ X_k^{-4,-3}, X_k^{-4,-2}, X_k^{-4,-1}, X_k^{-4,0}, X_k^{-4,1}, X_k^{-4,2}, X_k^{-4,3} \\ X_k^{-5,-4}, X_k^{-5,-3}, X_k^{-5,-2}, X_k^{-5,-1}, X_k^{-5,0}, X_k^{-5,1}, X_k^{-5,2}, X_k^{-5,3}, X_k^{-5,4} \end{array}$$

Table 4.1 Recursions implemented using four basic recursion formulae

Application	Direction	Initial values		Auxiliary recursion	Singularity
		One line	Two lines		
Ordinary Hansen coefficient	Forward of n	(R2)	(R1) and (R3)	(R4)	(R1):$m = 0$
	Backward of n	Lack	(R1)		
Eccentricity function	Forward of n		(R1) and (R3)	Lack	(R1):$m = 0$
	Backward of n		(R1)		

If we use basic recursion formula Eq. (R2) to perform the forward recursion of Hansen coefficients $X_k^{-(n+1),m}$ in the direction of n, the value range of the index m is $[-n, +n]$. Starting from $n = 2$ and progressing to the second line, only the three functions in the middle can be recursively derived, whereas the other four functions at either end need to be recursively calculated using basic recursion formula Eq. (R4) (or derived adopting direct calculation methods). Further down, there remain four functions at either end that cannot be recursively derived, and this is true for each line as the recursion proceeds. The same problem exists for the forward recursion in the direction of n when using basic recursion formulae Eqs. (R1) and (R3). This problem needs to be solved in the forward recursion in the direction of n, whereas there is no such problem in the backward recursion in the direction of n.

The recursions implemented using basic recursion formulae Eqs. (R1)–(R4) are summarized in Table 4.1. It is seen that there is no backward recursion formula of ordinary Hansen coefficients with one line of initial values, and there is no auxiliary recursion formula of the forward recursion of eccentricity functions.

4.2.3 Supplement to the Use of Basic Recursion Formulae

(1) **Backward recursion formula of ordinary Hansen coefficients with one line of initial values**

Using the basic recursion formulae Eqs. (R1) and (R2), it is straightforward to deduce a backward recursion formula of ordinary Hansen coefficients with one line of initial values. The specific deduction method is as follows.

Multiplying basic recursion formula Eq. (R1) by $\sqrt{1-e^2}$, we obtain

$$2m\left(1-e^2\right)X_k^{-(n+2),m} = ne\left[X_k^{-(n+1),m-1} - X_k^{-(n+1),m+1}\right] + 2k\sqrt{1-e^2}X_k^{-n,m}$$

Replacing n by $n+1$ ($n \Rightarrow n+1$) in basic recursion formula Eq. (R2) and multiplying by m, we have

$$2m\left(1-e^2\right)X_k^{-(n+2),m} = 2mX_k^{-(n+1),m} + me\left[X_k^{-(n+1),m+1} + X_k^{-(n+1),m-1}\right]$$

Through the subtraction of the above two expressions, we immediately obtain

$$2k\sqrt{1-e^2}X_k^{-n,m} = 2mX_k^{-(n+1),m} + (m+n)eX_k^{-(n+1),m+1} + (m-n)eX_k^{-(n+1),m-1} \tag{R5}$$

This is the backward recursion formula of ordinary Hansen coefficients with one line of initial values that we need [4].

From recursion formula Eq. (R5), we obtain one Hansen coefficient from three successive Hansen coefficient in a backward recursion. The recursion implemented using Eq. (R5) is illustrated below.

$$\begin{array}{ccc} & X_k^{-n,m} & \\ X_k^{-(n+1),m-1} & X_k^{-(n+1),m} & X_k^{-(n+1),m+1} \end{array} \qquad X_k^{-n,m} \Leftarrow X_k^{-(n+1),m-1}, X_k^{-(n+1),m}, X_k^{-(n+1),m+1}$$

(2) **Auxiliary recursion formula of the forward recursion of eccentricity functions**

Generally, basic recursion formulae Eqs. (R1) and (R3) are used to perform the forward recursion of eccentricity functions. We examine the following forward recursion process in the direction of n.

$$\begin{array}{ccccccccc} & & X_k^{-3,-2}, & \cancel{X_k^{-3,-1}}, & X_k^{-3,0}, & \cancel{X_k^{-3,1}}, & X_k^{-3,2} & & \\ & X_k^{-4,-3}, & \cancel{X_k^{-4,-2}}, & X_k^{-4,-1}, & \cancel{X_k^{-4,0}}, & X_k^{-4,1}, & \cancel{X_k^{-4,2}}, & X_k^{-4,3} & \\ X_k^{-5,-4}, & \cancel{X_k^{-5,-3}}, & X_k^{-5,-2}, & \cancel{X_k^{-5,-1}}, & X_k^{-5,0}, & \cancel{X_k^{-5,1}}, & X_k^{-5,2}, & \cancel{X_k^{-5,3}}, & X_k^{-5,4} \end{array}$$

Here, all functions not crossed out are eccentricity functions, and we expect to perform the recursion for these functions. From the known initial values (in the first and second lines), the recursion is performed to the third line, and the eccentricity functions in the middle are recursively processed. However, $X_k^{-5,-4}$ and $X_k^{-5,4}$ at either end cannot be recursively derived. A new recursion formula is thus needed.

For ordinary Hansen coefficients (including all functions in the above recursion), the recursion can be performed using basic recursion formula Eq. (R4). However, eccentricity functions do not appear successively in one line; i.e., eccentricity functions and ordinary Hansen coefficients appear alternately. As ordinary Hansen coefficients do not participate in the recursion, a new recursion formula must be deduced.

Without losing generality, we consider five eccentricity functions:

$$\begin{array}{c} X_k^{-n,m-3},\ X_k^{-n,m-1} \\ X_k^{-(n+1),m-4},\ X_k^{-(n+1),m-2},\ X_k^{-(n+1),m} \end{array}$$

The problem will be solved if we can deduce the recursion relation of these five functions.

Next, using the two basic recursion formulae Eqs. (R4) and (R5), we deduce the auxiliary recursion formula of the forward recursion of eccentricity functions. The specific deduction method is as follows.

Replacing m by $m+1$ ($m \Rightarrow m+1$) in basic recursion formulae Eq. (R5) and multiplying by $(m-1)$, we obtain

$$\begin{aligned}&2(m+1)(m-1)X_k^{-(n+1),m+1} = 2(m-1)k\sqrt{1-e^2}X_k^{-n,m+1}\\&-(m-1)(m+n+1)eX_k^{-(n+1),m+2} - (m-1)(m-n+1)eX_k^{-(n+1),m}\end{aligned} \tag{4.35}$$

Replacing m by $m-1$ ($m \Rightarrow m-1$) in basic recursion formula Eq. (R5) and multiplying by $(m+1)$, we obtain

$$\begin{aligned}&2(m-1)(m+1)X_k^{-(n+1),m-1} = 2k(m+1)\sqrt{1-e^2}X_k^{-n,m-1}\\&-(m+1)(m+n-1)eX_k^{-(n+1),m} - (m+1)(m-n-1)eX_k^{-(n+1),m-2}\end{aligned} \tag{4.36}$$

Substituting Eqs. (4.35) and (4.36) into basic recursion formula Eq. (R4) and multiplying by $(m+1)(m-1)$, we immediately obtain

$$\begin{aligned}&(m+1)(m-n-1)(n-m)e^2X_k^{-(n+1),m-2}\\&-(m+1)(m+n-1)(2m-n-1)e^2X_k^{-(n+1),m}\\&-(m-1)(m-n+1)(n+2m+1)e^2X_k^{-(n+1),m}\\&+(m+1)(m-1)\left[4m+2me^2-4k\left(1-e^2\right)^{3/2}\right]X_k^{-(n+1),m}\\&+2k(m-1)(n+2m+1)e\sqrt{1-e^2}X_k^{-n,m+1}\\&+2k(m+1)(2m-n-1)e\sqrt{1-e^2}X_k^{-n,m-1}\\&-(m-1)(n+m+1)(n+m)e^2X_k^{-(n+1),m+2}=0\end{aligned} \tag{R6}$$

This is the recursion formula that we need [4].

In particular, when $m=n-2$, we obtain $X_k^{-(n+1),n}$ from Eq. (R6):

$$\begin{aligned}&(n-3)(2n-1)e^2X_k^{-(n+1),n} = -3e^2X_k^{-(n+1),n-4}\\&+\left[2(n-3)(n-2)+3(n-2)e^2-2(n-3)k\left(1-e^2\right)^{3/2}\right]X_k^{-(n+1),n-2}\\&+3(n-3)ke\sqrt{1-e^2}X_k^{-n,n-1}+(n-5)ke\sqrt{1-e^2}X_k^{-n,n-3}\end{aligned} \tag{4.37}$$

When $m=-n+2$, we obtain $X_k^{-(n+1),-n}$ from Eq. (R6):

$$\begin{aligned}&(n-3)(2n-1)e^2X_k^{-(n+1),-n} = -3e^2X_k^{-(n+1),-n+4}\\&+\left[2(n-2)(n-3)+3(n-2)e^2+2(n-3)k\left(1-e^2\right)^{3/2}\right]X_k^{-(n+1),-n+2}\end{aligned}$$

$$- 3(n-3)ke\sqrt{1-e^2}X_k^{-n,-n+1} - (n-5)ke\sqrt{1-e^2}X_k^{-n,-n+3} \quad (4.38)$$

Equations (4.37) and (4.38) are the auxiliary recursion formulae of the forward recursion of eccentricity functions that we need.

4.2.4 Recursion of Ordinary Hansen Coefficients

The recursion discussed in this section can be applied to any Hansen coefficients $X_k^{-(n+1),m}$. According to the change in the index n, the recursion can be divided into forward recursion (increasing n) and backward recursion (decreasing n).

(1) **Forward recursion method for Hansen coefficients with Eq.** (R2)

Basic recursion formula Eq. (R2) is simple, and the forward recursion of Hansen coefficients $X_k^{-(n+1),m}$ using Eq. (R2) is the simplest. Equation (R2) requires only one line of initial values, and the recursion process is as follows.

$$X_k^{-3,-2},\ X_k^{-3,-1},\ X_k^{-3,0},\ X_k^{-3,1},\ X_k^{-3,2}$$
$$X_k^{-4,-3},\ X_k^{-4,-2},\ X_k^{-4,-1},\ X_k^{-4,0},\ X_k^{-4,1},\ X_k^{-4,2},\ X_k^{-4,3}$$
$$X_k^{-5,-4},\ X_k^{-5,-3},\ X_k^{-5,-2},\ X_k^{-5,-1},\ X_k^{-5,0},\ X_k^{-5,1},\ X_k^{-5,2},\ X_k^{-5,3},\ X_k^{-5,4}$$
$$\cdots\cdots$$
$$X_k^{-(n+1),-n},\ X_k^{-(n+1),-n+1},\ \cdots\ \cdots\ \cdots\ \cdots\ \cdots\ \cdots\ \cdots\ \cdots\ \cdots\ X_k^{-(n+1),n-1},\ X_k^{-(n+1),n}$$

Here, the first line comprises the initial values ($n = 2$), which must be calculated adopting a direct method of calculating Hansen coefficients (such as Wnuk's method [5]) in Chap. 3 in advance. As stated in Sect. 4.2.2, three functions in the middle of the second line can be recursively derived with Eq. (R2), whereas the other four functions at either end need to be 'calculated additionally'. The five functions in the middle of the third line can be recursively derived with Eq. (R2), whereas the other four functions at either end also need to be 'calculated additionally'. In other words, for each line of the recursion, there are four functions at either end that need to be calculated separately.

There is no singularity in the forward recursion of Hansen coefficients with Eq. (R2), which is a type of secure recursion. Moreover, the formula is simple, and the programming is convenient. The main disadvantage of this recursion method is that it requires direct methods of calculating Hansen coefficients (such as Wnuk's method [5]) to calculate the four functions at either end of each line of the recursion, which, of course, can also be achieved by recursion calculation with basic recursion formula Eq. (R4). However, there is a denominator of e in the recursion of Eq. (R4), which is unfavorable for small orbital eccentricities.

(2) **Backward recursion method of Hansen coefficients with Eq.** (R5)

One line of initial values is required to perform the backward recursion of Hansen coefficients $X_k^{-(n+1),m}$ using recursion formula Eq. (R5). The recursion process is as follows.

$$
\begin{array}{c}
X_k^{-(n+1),-n},\ X_k^{-(n+1),-n+1},\ \cdots\ \cdots\ \cdots\ \cdots\ \cdots\ \cdots\ \cdots\ \cdots\ \cdots\ X_k^{-(n+1),n-1},\ X_k^{-(n+1),n} \\
\cdots\cdots \\
X_k^{-5,-4},\ X_k^{-5,-3},\ X_k^{-5,-2},\ X_k^{-5,-1},\ X_k^{-5,0},\ X_k^{-5,1},\ X_k^{-5,2},\ X_k^{-5,3},\ X_k^{-5,4} \\
X_k^{-4,-3},\ X_k^{-4,-2},\ X_k^{-4,-1},\ X_k^{-4,0},\ X_k^{-4,1},\ X_k^{-4,2},\ X_k^{-4,3} \\
X_k^{-3,-2},\ X_k^{-3,-1},\ X_k^{-3,0},\ X_k^{-3,1},\ X_k^{-3,2}
\end{array}
$$

All Hansen coefficients can be recursively derived through this backward recursion, and there are no 'additional calculation' functions as in the forward recursion. Moreover, there is no singularity or e denominator. This is a type of secure recursion from the formula, but more initial values are required.

Incidentally, using Eqs. (3.74a) and (3.74b) in Chap. 3, the derivatives of Hansen coefficients can be calculated using the Hansen coefficients, once they are recursively derived. In this sense, the 'useless' functions of redundant calculations cannot be considered to be completely useless.

(3) **Forward recursion method of the kernels of Hansen coefficients and their derivatives with Eq.** (R2)

The kernels of Hansen coefficients are needed for the perturbation calculations without singularities:

$$K_k^{-(n+1),m} = e^{-|k-m|} X_k^{-(n+1),m} \tag{4.39}$$

Moreover, the derivatives of the kernels $K_k^{-(n+1),m}$ of Hansen coefficients are needed for the perturbation calculations without singularities. In the following, recursion methods of calculating the kernels $K_k^{-(n+1),m}$ of Hansen coefficients and their derivatives simultaneously are presented.

Substituting Eqs. (4.39) into basic recursion formula Eq. (R2) yields the recursion formula

$$
\begin{aligned}
\left(1-e^2\right)e^{|k-m|}K_k^{-(n+1),m} &= e^{|k-m|}K_k^{-n,m} \\
&\quad + \frac{e}{2}\left(e^{|k-m-1|}K_k^{-n,m+1} + e^{|k-m+1|}K_k^{-n,m-1}\right)
\end{aligned} \tag{4.40}
$$

We consider three recursion situations.

For $k-m=0$,

$$\left(1-e^2\right)K_k^{-(n+1),m} = K_k^{-n,m} + \frac{e^2}{2}\left(K_k^{-n,m+1} + K_k^{-n,m-1}\right) \tag{4.41a}$$

For $k-m \le -1$,

$$\left(1-e^2\right)K_k^{-(n+1),m} = K_k^{-n,m} + \frac{1}{2}\left(e^2 K_k^{-n,m+1} + K_k^{-n,m-1}\right) \tag{4.41b}$$

For $k-m \ge 1$,

$$\left(1-e^2\right)K_k^{-(n+1),m}=K_k^{-n,m}+\frac{1}{2}\left(K_k^{-n,m+1}+e^2K_k^{-n,m-1}\right) \tag{4.41c}$$

By differentiating Eqs. (4.41a)–(4.41c) with respect to e^2 and letting $K_k^{'-(n+1),m}=\frac{dK_k^{-(n+1),m}}{de^2}$, we have the following.

For $k-m=0$,

$$\begin{aligned}\left(1-e^2\right)K_k^{'-(n+1),m}&=K_k^{'-n,m}+\frac{e^2}{2}\left(K_k^{'-n,m+1}+K_k^{'-n,m-1}\right)\\&+\frac{1}{2}\left(K_k^{-n,m+1}+K_k^{-n,m-1}\right)+K_k^{-(n+1),m}\end{aligned} \tag{4.42a}$$

For $k-m\leq -1$,

$$\begin{aligned}\left(1-e^2\right)K_k^{'-(n+1),m}&=K_k^{'-n,m}+\frac{1}{2}\left(e^2K_k^{'-n,m+1}+K_k^{'-n,m-1}\right)\\&+\frac{1}{2}K_k^{-n,m+1}+K_k^{-(n+1),m}\end{aligned} \tag{4.42b}$$

For $k-m\geq 1$,

$$\begin{aligned}\left(1-e^2\right)K_k^{'-(n+1),m}&=K_k^{'-n,m}+\frac{1}{2}\left(K_k^{'-n,m+1}+e^2K_k^{'-n,m-1}\right)\\&+\frac{1}{2}K_k^{-n,m-1}+K_k^{-(n+1),m}\end{aligned} \tag{4.42c}$$

Equations (4.41a, 4.41b, 4.41c) and (4.42a, 4.42b, 4.42c) are the recursion formulae for calculating the kernels $K_k^{-(n+1),m}$ of Hansen coefficients and their derivatives $K_k^{'-(n+1),m}=\frac{dK_k^{-(n+1),m}}{de^2}$ with respect to e^2. Using the method described in Chap. 3 to calculate the initial values of $K_k^{-3,m}$ and $K_k^{'-3,m}$ $(m=-2,-1,\cdots,2)$, $K_k^{-4,m}$ and $K_k^{'-4,m}$ $(m=-1,0,1)$ are recursively derived with Eqs. (4.41a, 4.41b, 4.41c) and (4.42a, 4.42b, 4.42c). Furthermore, $K_k^{-4,m}$ and $K_k^{'-4,m}(m=-3,-2,2,3)$ are calculated through the recursion of Eq. (R4) (or a direct calculation method). Continuing to perform the recursions using Eqs. (4.41a, 4.41b, 4.41c) and (4.42a, 4.42b, 4.42c) step by step, $K_k^{-(n+1),m}$ and $K_k^{'-(n+1),m}$ $(m=-n+2,-n+3,\cdots,n-2)$ are recursively derived. Using the recursion of Eq. (R4) (or the direct calculation methods), $K_k^{-(n+1),m}$ and $K_k^{'-(n+1),m}(m=-n,-n+1,n-1,n)$ are also calculated. All $K_k^{-(n+1),m}$ and $K_k^{'-(n+1),m}$ are then derived eventually.

Of course, here, $K_k^{'-(n+1),m}$ denotes derivatives with respect to e^2, which should be converted into derivatives with respect to e through $\frac{dK_k^{-(n+1),m}}{de}=2e\frac{dK_k^{-(n+1),m}}{de^2}=2eK_k^{'-(n+1),m}$ when used. This forward recursion method still has the disadvantage of 'additional calculations'. If the forward recursion is changed into the backward recursion, the disadvantage of 'additional calculations' can be avoided.

(4) **Backward recursion method of the kernels of Hansen coefficients with Eq. (R5)**

Substituting Eq. (4.39) into basic recursion formula Eq. (R5) gives the recursion formula

$$\begin{aligned}2k\sqrt{1-e^2}e^{|k-m|}K_k^{-n,m} &= 2me^{|k-m|}K_k^{-(n+1),m}\\ &+ (m+n)e^{|k-m-1|+1}K_k^{-(n+1),m+1} + (m-n)e^{|k-m+1|+1}K_k^{-(n+1),m-1}\end{aligned} \tag{4.43}$$

Again, we consider three recursion situations.

For $k-m=0$,

$$\begin{aligned}2k\sqrt{1-e^2}K_k^{-n,m} &= 2mK_k^{-(n+1),m} + (m+n)e^2K_k^{-(n+1),m+1}\\ &+ (m-n)e^2K_k^{-(n+1),m-1}\end{aligned} \tag{4.44a}$$

For $k-m\le -1$,

$$2k\sqrt{1-e^2}K_k^{-n,m} = 2mK_k^{-(n+1),m} + (m+n)e^2K_k^{-(n+1),m+1} + (m-n)K_k^{-(n+1),m-1} \tag{4.44b}$$

For $k-m\ge 1$,

$$2k\sqrt{1-e^2}K_k^{-n,m} = 2mK_k^{-(n+1),m} + (m+n)K_k^{-(n+1),m+1} + (m-n)e^2K_k^{-(n+1),m-1} \tag{4.44c}$$

4.2.5 *Recursion of Eccentricity Functions*

(1) **Vakhidov's recursion method**

Vakhidov proposed a recursion method whereby only useful Hansen coefficients in the perturbations are calculated [3]; i.e., the recursion method of eccentricity functions. Vakhidov deduced formulae adopting computer algebra owing to their complexity. To clarify the deduction process, the central idea of Vakhidov's method is introduced as follows.

The main recursion formula of Vakhidov's method is the basic recursion formula Eq. (R1), and the basic recursion formula Eq. (R3) is needed when $m=0$. The initial values of the recursion are eccentricity functions $X_k^{-(n+1),m}$ of the orders $n=2$ and $n=3$. The recursion process of Vakhidov's method is as follows.

$$
\begin{array}{c}
X_k^{-3,-2},\ \cancel{X_k^{-3,-1}},\ X_k^{-3,0},\ \cancel{X_k^{-3,1}},\ X_k^{-3,2} \\
X_k^{-4,-3},\ \cancel{X_k^{-4,-2}},\ X_k^{-4,-1},\ \cancel{X_k^{-4,0}},\ X_k^{-4,1},\ \cancel{X_k^{-4,2}},\ X_k^{-4,3} \\
X_k^{-5,-4},\ \cancel{X_k^{-5,-3}},\ X_k^{-5,-2},\ \cancel{X_k^{-5,-1}},\ X_k^{-5,0},\ \cancel{X_k^{-5,1}},\ X_k^{-5,2},\ \cancel{X_k^{-5,3}},\ X_k^{-5,4} \\
\cdots\cdots \\
X_k^{-(n+1),-n},\ \cancel{X_k^{-(n+1),-n+1}},\ \cdots\ \cdots\ \cdots\ \cdots\ \cdots\ \cdots\ \cdots\ \cdots\ \cdots\ \cancel{X_k^{-(n+1),n-1}},\ X_k^{-(n+1),n}
\end{array}
$$

Here, the functions not crossed out are eccentricity functions needed in the perturbation calculations, whereas the functions crossed out are useless in the perturbation calculations. In the recursion, the eccentricity functions in the first two lines are required to recursively derive those in the third line step by step. It is not difficult to see that the eccentricity functions in the middle of each line can be recursively derived using the basic recursion formulae Eqs. (R1) and (R3). However, the eccentricity functions at either end cannot be recursively derived. Therefore, the key to solving the recursion of eccentricity functions is deducing the recursion formula for calculating eccentricity functions at either end.

The formula for calculating $X_k^{-(n+1),-n}$ presented by Vakhidov [3] is

$$
\alpha_1 X_s^{n-2,m-3} = \beta_1 X_s^{n-1,m-2} + \gamma_1 X_s^{n-1,m} + \delta_1 X_s^{n-1,m+2} + \varepsilon_1 X_s^{n,m-1} + \varphi_1 X_s^{n,m+1} \tag{V1}
$$

where $s \neq 0$, $n < 0$ and

$$
\begin{aligned}
\alpha_1 &= 2e^2\left(1-e^2\right)^{3/2}(m+n-2)(m+n-3)\left(m^2-1\right) \\
\beta_1 &= 4se\left(2m^3+2m^2n-4m^2-nm-3n-2m+4\right)\left(1-e^2\right)^2 \\
&\quad + 2ne\left(m^2-1\right)\sqrt{1-e^2}\left[e^2(2m+2n-3)-2(m-2)\right] \\
\gamma_1 &= 4se(m+1)\left(n+2m^2-6m+4\right)\left(1-e^2\right)^2 \\
&\quad + e^3n\sqrt{1-e^2}\left(3n-n^2m+2-3m-2m^2+3m^3-nm-2m^2n+n^2\right) \\
&\quad + 4e\sqrt{1-e^2}mn\left(1-m^2\right) \\
\delta_1 &= e^3\sqrt{1-e^2}n(m-1)(m-n)(m-n+1) \\
\varepsilon_1 &= 8s^2\left(1-e^2\right)^{5/2}\left(2-m^2+m\right) + 4se^4\left(m^2-1\right)(m+2n-2) \\
&\quad + 4se^2\left(1-m^2\right)(2n+3m-6) + 8s\left(m^2-1\right)(m-2) \\
\varphi_1 &= 2se^2\left(1-e^2\right)\left(-m^3+2m^2+mn^2-n^2-2m^2n+nm+n+m-2\right)
\end{aligned}
$$

The formula for calculating $X_k^{-(n+1),n}$ presented by Vakhidov [3] is

$$
\alpha_2 X_s^{n-2,m+3} = \beta_2 X_s^{n-1,m+2} + \gamma_2 X_s^{n-1,m} + \delta_2 X_s^{n-1,m-2} + \varepsilon_2 X_s^{n,m+1} + \varphi_2 X_s^{n,m-1} \tag{V2}
$$

where $s \neq 0$, $n < 0$ and

$$\alpha_2 = 2e^2\left(1-e^2\right)^{3/2}(m-n+2)(m-n+3)\left(m^2-1\right)$$
$$\beta_2 = 4se\left(2m^3-2m^2n+4m^2-nm+3n-2m-4\right)\left(1-e^2\right)^2$$
$$+ 2ne\left(m^2-1\right)\sqrt{1-e^2}\left[e^2(2n-2m-3)+2(m+2)\right]$$
$$\gamma_2 = 4se(m-1)\left(n+2m^2+6m+4\right)\left(1-e^2\right)^2$$
$$+ e^3n\sqrt{1-e^2}\left(3n+n^2m+2+3m-2m^2-3m^3+nm-2m^2n+n^2\right)$$
$$+ 4e\sqrt{1-e^2}mn\left(m^2-1\right)$$
$$\delta_2 = e^3\sqrt{1-e^2}n(m+1)(m+n)(1-m-n)$$
$$\varepsilon_2 = 8s^2\left(1-e^2\right)^{5/2}\left(2-m^2-m\right)+4se^4\left(m^2-1\right)(m-2n+2)$$
$$+ 4se^2\left(m^2-1\right)(2n-3m-6)+8s\left(m^2-1\right)(m+2)$$
$$\varphi_2 = 2se^2\left(1-e^2\right)\left(-m^3-2m^2+mn^2+n^2+2m^2n+nm-n+m+2\right)$$

When replacing s and n in Eqs. (V1) and (V2) with k and $-n$, respectively ($s \Rightarrow k$ and $n \Rightarrow -n$), the forms of the corresponding functions are consistent with those in this monograph.

Using Vakhidov's recursion method, for the given indexes k and N (i.e., the upper limit of the index n), two lines of initial values are presented as follows.

$$X_k^{-3,-2},\ X_k^{-3,0},\ X_k^{-3,2}$$
$$X_k^{-4,-3},\ X_k^{-4,-1},\ X_k^{-4,1},\ X_k^{-4,3}$$

Then, $X_k^{-(n+1),m}$ $(n = 4, 5, \cdots, N; m = -n, -n+2, \cdots, n-2, n)$ can be recursively derived.

It is seen that the main contributions of Vakhidov's recursion method are that the recursion formulae Eqs. (V1) and (V2) are deduced using computer algebra such that the recursion can be performed among the 'useful' eccentricity functions. That is to say, the right sides of the recursion formulae are all 'useful' eccentricity functions, and the functions derived recursively are all 'useful' eccentricity functions, which saves much memory and computing time.

(2) **Forward recursion method of eccentricity functions with Eqs.** (R1), (R3) and (R6)

As Vakhidov's recursion formulae Eqs. (V1) and (V2) are deduced through computer algebra, the formulae are obviously complex, and improvements to the methodology should be made. Our improvement of the methodology is to use recursion formula Eq. (R6) and combine basic recursion formulae Eqs. (R1) and (R3) to perform the corresponding forward recursion of eccentricity functions in the direction of n. The recursion process is the same as that in Vakhidov's method. The eccentricity functions in the middle of each line are first derived recursively using basic recursion formula Eq. (R1) (when $m = 0$, the basic recursion formula Eq. (R3) is used to perform the recursion) and the eccentricity functions $X_k^{-(n+1),n}$ and

$X_k^{-(n+1),-n}$ at either end are then calculated using recursion formulae Eqs. (4.37) and (4.38).

Similar to Vakhidov's recursion method, for given indexes k and N (i.e., the upper limit of the index n), the forward recursion of eccentricity functions can be done once the two lines of initial values are presented. The right sides of recursion formulae Eqs. (4.37) and (4.38) are all 'useful' eccentricity functions, which can also achieve the purpose for recursion among eccentricity functions. Moreover, the recursion formulae Eqs. (4.37) and (4.38) are much simpler than the recursion formulae Eqs. (V1) and (V2). Furthermore, there is no singularity in the recursion formulae Eqs. (4.37) and (4.38), but there are e denominators.

(3) **Backward recursion method of eccentricity functions with Eq.** (R1)

In fact, Vakhidov's recursion method adopts forward recursion in the direction of n with the basic recursion formulae Eqs. (R1) and (R3). There is a problem with conducting the 'additional calculations' of $X_k^{-(n+1),n}$ and $X_k^{-(n+1),-n}$ in this way. If the backward recursion in the direction of n using the basic recursion formulae Eq. (R1) is adopted, this problem does not exist, and the corresponding recursion formula is

$$kX_k^{-n,m} = m\sqrt{1-e^2}X_k^{-(n+2),m} - \frac{ne}{2\sqrt{1-e^2}}\left[X_k^{-(n+1),m-1} - X_k^{-(n+1),m+1}\right] \quad (4.45)$$

The backward recursion with the recursion formula Eq. (4.45) needs to calculate two lines of initial values, but it is applicable to the recursion of eccentricity functions. After eccentricity functions are recursively derived, their derivatives can be calculated according to [1]

$$\frac{dX_k^{n,m}}{de} = \frac{2m-n}{2}X_k^{n-1,m+1} - \frac{2m+n}{2}X_k^{n-1,m-1} - \frac{me}{4(1-e^2)}\left(X_k^{n,m+2} - X_k^{n,m-2}\right) \quad (4.46)$$

For the given indexes k and N (i.e., the upper limit of the index n), two lines of initial values of eccentricity functions are presented:

$$X_k^{-(N+1),m} \quad (m = -N, -N+2, \cdots, N-2, N)$$

$$X_k^{-N,m} \quad (m = -N+1, -N+3, \cdots, N-3, N-1)$$

Then, using recursion formula Eq. (4.45), these eccentricity functions are recursively derived immediately as

$$X_k^{-(n+1),m} \quad (n = N-2, N-3, \cdots, 2; m = -n, -n+2, \cdots, n-2, n)$$

This method is simple to implement, and there is no singularity or denominator of e. It might therefore be a relatively appropriate method for the recursion of eccentricity functions.

(4) **Backward recursion method of the kernels of eccentricity functions and their derivatives with Eq.** (R1)

If the kernels of eccentricity functions need to be recursively derived, $X_k^{-(n+1),m} = e^{|k-m|} K_k^{-(n+1),m}$ is substituted into recursion formula Eq. (4.45) to obtain

$$\begin{aligned} k e^{|k-m|} K_k^{-n,m} &= m\sqrt{1-e^2}\, e^{|k-m|} K_k^{-(n+2),m} \\ &\quad - \frac{ne}{2\sqrt{1-e^2}}\left[e^{|k-m+1|} K_k^{-(n+1),m-1} - e^{|k-m-1|} K_k^{-(n+1),m+1}\right] \end{aligned} \tag{4.47}$$

Here, we consider three recursion situations.

For $k - m \geq 1$,

$$kK_k^{-n,m} = m\sqrt{1-e^2} K_k^{-(n+2),m} - \frac{n}{2\sqrt{1-e^2}}\left[e^2 K_k^{-(n+1),m-1} - K_k^{-(n+1),m+1}\right] \tag{4.48a}$$

For $k - m = 0$,

$$kK_k^{-n,m} = m\sqrt{1-e^2} K_k^{-(n+2),m} - \frac{ne^2}{2\sqrt{1-e^2}}\left[K_k^{-(n+1),m-1} - K_k^{-(n+1),m+1}\right] \tag{4.48b}$$

For $k - m \leq -1$,

$$kK_k^{-n,m} = m\sqrt{1-e^2} K_k^{-(n+2),m} - \frac{n}{2\sqrt{1-e^2}}\left[K_k^{-(n+1),m-1} - e^2 K_k^{-(n+1),m+1}\right] \tag{4.48c}$$

By differentiating Eqs. (4.48a)–(4.48c) with respect to e^2 and letting $K_k^{'-(n+1),m} = \frac{dK_k^{-(n+1),m}}{de^2}$, we have the following.

For $k - m \geq 1$,

$$\begin{aligned} k\sqrt{1-e^2} K_k^{'-n,m} &= m\left(1-e^2\right) K_k^{'-(n+2),m} - \frac{n}{2}\left[e^2 K_k^{'-(n+1),m-1} - K_k^{'-(n+1),m+1}\right] \\ &\quad - mK_k^{-(n+2),m} + \frac{k}{2\sqrt{1-e^2}} K_k^{-n,m} - \frac{n}{2} K_k^{-(n+1),m-1} \end{aligned} \tag{4.49a}$$

For $k - m = 0$,

$$k\sqrt{1-e^2} K_k^{'-n,m} = m\left(1-e^2\right) K_k^{'-(n+2),m} - \frac{ne^2}{2}\left[K_k^{'-(n+1),m-1} - K_k^{'-(n+1),m+1}\right]$$

$$-mK_k^{-(n+2),m}+\frac{k}{2\sqrt{1-e^2}}K_k^{-n,m}$$
$$-\frac{n}{2}\left[K_k^{-(n+1),m-1}-K_k^{-(n+1),m+1}\right] \quad (4.49b)$$

For $k-m\leq -1$,

$$k\sqrt{1-e^2}K_k^{'-n,m}=m\left(1-e^2\right)K_k^{'-(n+2),m}-\frac{n}{2}\left[K_k^{'-(n+1),m-1}-e^2K_k^{'-(n+1),m+1}\right]$$
$$-mK_k^{-(n+2),m}+\frac{k}{2\sqrt{1-e^2}}K_k^{-n,m}+\frac{n}{2}K_k^{-(n+1),m+1} \quad (4.49c)$$

For the given indexes k and N (i.e., the upper limit of the index n), two lines of initial values of the kernels of eccentricity functions and their derivatives $K_k^{'-(n+1),m}=\frac{dK_k^{-(n+1),m}}{de^2}$ with respect to e^2 are presented:

$$K_k^{-(N+1),m}\ (m=-N,-N+2,\cdots,N-2,N)$$

$$K_k^{-N,m}\ (m=-N+1,-N+3,\cdots,N-3,N-1)$$

$$K_k^{'-(N+1),m}=\frac{dK_k^{-(N+1),m}}{de^2}\quad (m=-N,-N+2,\cdots,N-2,N)$$

$$K_k^{'-N,m}=\frac{dK_k^{-N,m}}{de^2}\quad (m=-N+1,-N+3,\cdots,N-3,N-1)$$

Using recursion formulae Eqs. (4.48a, 4.48b, 4.48c) and (4.49a, 4.49b, 4.49c), these kernels of eccentricity functions and their derivatives with respect to e^2 are recursively derived immediately:

$$K_k^{-(n+1),m},\ K_k^{'-(n+1),m}$$
$$=\frac{dK_k^{-(n+1),m}}{de^2}\quad (n=N-2,N-3,\cdots,2;\ m=-n,-n+2,\cdots,n-2,n)$$

Then, according to $\frac{dK_k^{-(n+1),m}}{de}=2e\frac{dK_k^{-(n+1),m}}{de^2}=2eK_k^{'-(n+1),m}$, the derivatives of the kernels of eccentricity functions with respect to e are further obtained.

(5) **Batch calculation method for eccentricity functions and their derivatives**

The eccentricity functions and their derivatives can be batch calculated using the definite integral method of Hansen coefficients and their derivatives. The corresponding calculation formulae remain

$$X_k^{n,m}=\frac{1}{\pi}\int_0^{\pi}\left(\frac{r}{a}\right)^{n+1}\cos(mf-kM)dE$$

$$\frac{dX_k^{n,m}}{de} = -\frac{1}{\pi}\int_0^{\pi}\left(\frac{r}{a}\right)^{n+1}\left[n\frac{a}{r}\cos f\cos(mf - kM)\right.$$
$$\left. + \frac{m}{1-e^2}(2 + e\cos f)\sin f\sin(mf - kM)\right]dE$$

Using the definite integral calculation, eccentricity functions $G_{lpq} = X_{l-2p+q}^{-(l+1),l-2p}$ and their derivatives $\frac{dG_{lpq}}{de}$ are calculated at one time ($p = 0, 1, \cdots, l$, $q = -Q, -Q+1, \cdots, Q$, and Q is the maximum of the index q).

Of course, such a batch calculation of eccentricity functions and their derivatives is not a recursion method but a direct calculation method. However, in the batch calculation process, the quantities related to l or n (including $\frac{r}{a}$, f and M in the calculation formulae) are calculated only once, and the symmetry of eccentricity functions $G_{lpq} = G_{l,l-p,-q}$ can be used. For the index p, only the cases of $0, 1, \cdots, l/2$ are calculated, and the calculation quantity is reduced by half. As a result, the computational efficiency of this batch calculation method for eccentricity functions and their derivatives is relatively higher.

(6) **Batch recursion method for eccentricity functions**

The core of the batch recursion method for eccentricity functions remains a backward recursion with basic recursion formula Eq. (R1) and a forward recursion with basic recursion formulae Eqs. (R1) and (R3).

The backward recursion method of eccentricity functions with Eq. (R1) is first analyzed. Here, Eq. (4.4) is used as the recursion formula; i.e.,

$$(l-2p+q)\sqrt{1-e^2}G_{l-2,p-1,q} = (l-2p)\left(1-e^2\right)G_{lpq} + \frac{(l-1)e}{2}\left(G_{l-1,p-1,q-1} - G_{l-1,p,q+1}\right) \quad (4.50)$$

The initial values of the recursion are

$$G_{lpq}\ (p = 0, 1, \cdots, l; q = -Q, -Q+1, \cdots, Q)$$
$$G_{l-1,p,q}\ (p = 0, 1, \cdots, l-1; q = -Q, -Q+1, \cdots, Q)$$

where Q is the maximum of the index q. In the recursion, the values of the index p are set as $1, 2, \cdots, l-1$ (where the symmetry of eccentricity functions is not considered), and the eccentricity functions are then calculated as

$$G_{l-2,p-1,q}\ (p = 1, 2, \cdots, l-1; q = -Q, -Q+1, \cdots, Q)$$

All eccentricity functions G_{lpq} are recursively derived by cycling the index l; i.e., the values of l are taken from L to 4 (where L is the maximum of the index l). Of course, the recursion can be performed again and again. The perturbation associated with a value of l can be obtained for each recursion, enabling a reduction in data storage. However, such recursion is incomplete. Attention should be paid to two problems. (1) When $l - 2p + q = 0$, the recursion method of Hansen coefficients with the index $k = 0$ described in Chap. 2, i.e., the recursion method of $X_0^{-(n+1),m}$, is required. (2) When $q = \pm Q$, there are no initial values for the term $\left(G_{l-1,p-1,q-1} - G_{l-1,p,q+1}\right)$ in Eq. (4.50), and the direct calculation method of Hansen coefficients is required.

It is clear that there are eccentricity functions G_{lpq} for 3 values of q that must be 'calculated additionally' using the above recursion method. In other words, when $Q = 10$, the recursion cannot be performed for 3 out of 21 values of q, whereas there are still eccentricity functions G_{lpq} for 18 values of q that can be recursively derived. Using the symmetry of eccentricity functions $G_{lpq} = G_{l,l-p,-q}$, the cost of calculating eccentricity functions G_{lpq} can be reduced by half, and the computational efficiency is much higher than that of the recursion methods with the index k equal to a constant singly. Nevertheless, it is noted that there are obvious disadvantages of the batch recursion method of using recursion formula Eq. (4.50) to perform the backward recursion. On the one hand, the method requires initial values of the higher-order eccentricity functions, whose errors may be large when the eccentricity is large. On the other hand, it is easy to lose significant digits when the recursion is performed from the eccentricity functions with large magnitudes to those with small magnitudes.

To overcome the above obvious disadvantages, the forward recursion method with basic recursion formula Eq. (R3) is used. The corresponding recursion formula is

$$\begin{aligned}\left(1-e^2\right)^2 G_{lpq} = &\left(1-\frac{e^2}{2}\right)G_{l-2,p-1,q} + \left(1-e^2\right)e\left(G_{l-1,p-1,q-1} + G_{l-1,p,q+1}\right)\\ &- \frac{e^2}{4}\left(G_{l-2,p-2,q-2} + G_{l-2,p,q+2}\right)\end{aligned} \tag{4.51}$$

The index p must be $0, 1, \cdots, l$ to obtain eccentricity functions G_{lpq} for all indexes of p and q. There are then cases that p exceeds the limit. In addition, there are cases that q exceeds the limit as the above backward recursion method with recursion formula Eq. (4.50). The cases that the indexes p and q may exceed the limits are listed in Table 4.2. When using the forward recursion method with recursion formula Eq. (4.51), eccentricity functions G_{lpq} for $p = 0$, $p = 1$, $p = l - 1$, $p = l$, $q = -Q$, $q = -Q + 1$, $q = Q - 1$ and $q = Q$ must be calculated directly from Table 4.2 as they are unable to be derived from the recursion. According to the symmetry of eccentricity functions, the cases of exceeding the limit at $p = l - 1$ and $p = l$ can be avoided. There is no singularity or e denominator in the forward recursion of eccentricity functions with recursion formula Eq. (4.51), whose calculation quantity is larger than that of the backward recursion with recursion formula Eq. (4.50). However, the calculation quantity can be reduced by half by using the symmetry of eccentricity functions, and the method is thus acceptable.

Similarly, using the forward recursion with basic recursion formula Eq. (R1) in the case that $l - 2p \neq 0$, the related terms in the recursion formula Eq. (4.4) exceed limits at $p = 0$, $p = l$, and $q = \pm Q$. While using the forward recursion

Table 4.2 Cases that the indexes p and q exceed the limits in the batch recursion of eccentricity functions G_{lpq} using recursion formula Eq. (4.51) to perform the forward recursion

Term	The case that p exceed the limit	The case that q exceed the limit
$G_{l-2,p-1,q}$	$p = 0, p = l$	
$G_{l-1,p-1,q-1}$	$p = 0$	$q = -Q$
$G_{l-1,p,q+1}$	$p = l$	$q = Q$
$G_{l-2,p-2,q-2}$	$p = 0, p = 1$	$q = -Q, q = -Q + 1$
$G_{l-2,p,q+2}$	$p = l - 1, p = l$	$q = Q - 1, q = Q$

with basic recursion formula Eq. (R3) in the case that $l - 2p = 0$ (i.e., in the case that $p = l/2$) only, the related terms in the recursion formula Eq. (4.51) exceed limits at $q = \pm(Q - 1)$ and $q = \pm Q$ only. In this way, it can overcome the obvious disadvantages of backward recursion with basic recursion formula Eq. (R1), i.e., recursion formula Eq. (4.50), and reduce the calculation quantity of the forward recursion with basic recursion formula Eq. (R3), i.e., recursion formula Eq. (4.51). Supposing that the required highest order of the disturbing function of the Earth's gravitational field is 100 and the summation range of the index q of eccentricity functions G_{lpq} is -10 to 10, the total number of eccentricity functions G_{lpq} to be calculated is $5148 \times 21 = 108108$. In fact, only 54,079 eccentricity functions need to be calculated using the symmetry of eccentricity functions, and it is the calculation quantity of eccentricity functions G_{lpq} when using the direct calculation method. Using the symmetry of eccentricity functions, the calculation quantity $\tilde{Q}$ for eccentricity functions G_{lpq} in batch recursion using forward recursion methods with basic recursion formulae Eqs. (R1) and (R3), i.e., recursion formulae Eqs. (4.4) and (4.51), is

$$\begin{cases} \tilde{Q} = \tilde{Q}_d + \tilde{Q}_r \times \dfrac{T_r}{T_d} \\ \tilde{Q}_d = 21 + 11 + 2 \times 21 + 49 \times 2 + 97 \times 19 + \displaystyle\sum_{l=4}^{100} \left[\frac{l+1}{2}\right] \times 2 = 7107 \\ \tilde{Q}_r = 54079 - \tilde{Q}_d = 46972 \end{cases} \tag{4.52}$$

where $\tilde{Q}_d$ and $\tilde{Q}_r$ are the numbers of eccentricity functions derived adopting the direct calculation method and recursion methods, respectively, T_r/T_d is the ratio of the recursion calculation time to the direct calculation time, and $[\cdot]$ refers to taking the integer part of the term. Supposing $T_r/T_d = 0.01$, the calculation quantity of eccentricity functions G_{lpq} of the batch recursion is $\tilde{Q} \approx 7577$ from Eq. (4.52), which is merely 14.01% of that of the direct calculation method, and the corresponding computational efficiency is approximately 6.14 times higher than the direct calculation method. Supposing $T_r/T_d = 0.003$, the calculation quantity of eccentricity functions G_{lpq} of the batch recursion is $\tilde{Q} \approx 7248$ from Eq. (4.52), which is merely 13.40% of that of the direct calculation method, and the corresponding computational efficiency is approximately 6.46 times higher than the direct calculation method.

Thus, by adopting forward recursion methods with basic recursion formulae Eqs. (R1) and (R3), i.e., recursion formulae Eqs. (4.4) and (4.51), a batch recursion method having a high computational efficiency for eccentricity functions is introduced [6]. The corresponding recursion formula is

$$G_{lpq} = \begin{cases} \dfrac{1}{\left(1-e^2\right)^2}\Bigg[\left(1-\dfrac{e^2}{2}\right)G_{l-2,p-1,q} + \left(1-e^2\right)e\left(G_{l-1,p-1,q-1} + G_{l-1,p,q+1}\right) \\ -\dfrac{e^2}{4}\left(G_{l-2,p-2,q-2} + G_{l-2,p,q+2}\right)\Bigg], \quad (l-2p=0) \\ \dfrac{1}{(l-2p)\left(1-e^2\right)}\Bigg[(l-2p+q)\sqrt{1-e^2}G_{l-2,p-1,q} \\ -\dfrac{(l-1)e}{2}\left(G_{l-1,p-1,q-1} - G_{l-1,p,q+1}\right)\Bigg], \quad (l-2p \neq 0) \end{cases} \tag{4.53}$$

The most important advantages of this batch recursion method are that it is forward and requires fewer initial values and that the magnitudes of eccentricity functions undergo changes from small to large in the recursion process, such that high accuracy of the recursion of eccentricity functions is guaranteed.

4.3 Comparison of Computational Efficiency between the Direct Calculation Method and Recursion Methods of Hansen Coefficients

Many recursion methods of Hansen coefficients were introduced above. The use of recursion methods is known to improve computational efficiency. In the following, the computational efficiencies of the direct calculation method and recursion methods with the index k equal to a constant for Hansen coefficients $X_k^{-(n+1),m}$ are compared in a specific example. As stated above, supposing that the required highest order of the disturbing function of the Earth's gravitational field is 100 and the summation range of the index q of eccentricity functions G_{lpq} is -10 to 10, with the aid of symmetry of eccentricity functions, the calculation quantity of eccentricity functions G_{lpq} is 54079 when using the direct calculation method.

However, when recursion methods with the index $k = l - 2p + q$ equal to a constant of eccentricity functions $G_{lpq} = X_{l-2p+q}^{-(l+1),l-2p}$ are adopted, there need to be two lines of initial values for the forward recursion with recursion formulae Eqs. (R1), (R3), and (R6) and the backward recursion with recursion formula Eq. (R1). Accordingly, the calculation quantities $\tilde{Q}_f$ and $\tilde{Q}_b$ of eccentricity functions G_{lpq} in the forward recursion and backward recursion are respectively

$$\begin{cases} \tilde{Q}_f = 7 \times \tilde{K} + 108101 \times \tilde{K} \times \dfrac{T_r}{T_d} \\ \tilde{Q}_b = 201 \times \tilde{K} + 107907 \times \tilde{K} \times \dfrac{T_r}{T_d} \end{cases} \tag{4.54}$$

where $\tilde{K}$ is a constant determined by the three indexes l, p, and q of eccentricity functions G_{lpq}, T_r is the recursion calculation time, and T_d is the direct calculation time. According to the above same requirement as the direct calculation method, $\tilde{K} = 111$, considering the symmetry of eccentricity functions.

Supposing the ratio of the recursion calculation time to the direct calculation time $T_r / T_d = 0.01$, the calculation quantity of eccentricity functions G_{lpq} is $\tilde{Q}_f \approx 120769$ in the forward recursion from Eq. (4.54) and $\tilde{Q}_b \approx 142088$ in the backward recursion. This is clearly not cost-effective for the recursion calculation of eccentricity functions G_{lpq}. Even for the ratio of the recursion calculation time to the direct calculation time $T_r / T_d = 0.003$, the calculation quantity of eccentricity functions G_{lpq} is $\tilde{Q}_f \approx 36775$ in the forward recursion from Eq. (4.54) and $\tilde{Q}_b \approx 58244$ in the backward recursion. Thus, only the forward recursion is cost-effective for the recursion calculation of eccentricity functions G_{lpq}, and its computational efficiency is higher than that of the direct calculation method. In addition, even if the recursion calculation time $T_r = 0$, it is clear that the calculation quantity $\tilde{Q}_b$ of eccentricity functions G_{lpq} in the backward recursion is approximately 41.26% of that of the direct calculation method from Eq. (4.54), and the corresponding computational efficiency increases only approximately 1.42 times.

As stated in Sect. 4.2.5, if the batch recursion method of eccentricity functions is adopted, the calculation quantity is significantly less than that of the direct calculation method. The computational efficiency is then superior to that of the direct calculation method and much higher than that of the corresponding recursion methods with the index k equal to a constant. However, to improve the computational efficiency, it is necessary to further study a more efficient recursion method for Hansen coefficients.

4.4 Recursion Example

The calculation formulae of recursion methods are relatively simple, and hence only the backward recursion method of recursion formula Eq. (R5) and the forward recursion method of recursion formula Eq. (R6) are adopted in the recursion example. The backward recursion method of Eq. (R5) can be applied to the recursion of ordinary Hansen coefficients and it is the simplest. The forward recursion method of Eq. (R6) can be applied to the recursion of eccentricity functions. Three recursion formulae Eqs. (R1), (R3), and (R6) are involved actually and the calculation program is complex.

Table 4.3 presents calculation results for an index $k = 1$ and eccentricity $e = 0.1$. This example is taken to present the recursion processes of the recursion methods of Eqs. (R5) and (R6) clearly and to see which initial values are required and what results are recursively derived. Table 4.3 lists six columns of data, in which the first three columns are the indexes n, $-(n+1)$, and m of Hansen coefficients $X_k^{-(n+1),m}(e)$ and the last three columns are the calculation results of Wnuk's method [5], the backward recursion method of Eq. (R5), and the forward recursion method of Eq. (R6). Here, the calculation result of Wnuk's method is taken as a reference with which to compare the recursion results of Eqs. (R5) and (R6). In the last two columns of Table 4.3, the shaded data are the initial values, and the unshaded data are the recursion results.

For the backward recursion of Eq. (R5), all eleven data of $n = 5$ (or $-(n+1) = -6$) are required as initial values. Here, the recursion starts from $n = 5$ (or $-(n+1) = -6$). More initial values are required for the recursion of higher-order Hansen coefficients. For the forward recursion of Eq. (R6), seven data in two lines of $n = 2$ and $n = 3$ (or $-(n+1) = -3$ and $-(n+1) = -4$) are required as initial values, but only the eccentricity functions are needed. In addition, there are also only eccentricity functions in the recursion result.

Table 4.3 reveals that the backward recursion result of Eq. (R5) is good and consistent with the calculation result of Wnuk's method. However, there is an error at the beginning in the forward recursion of Eq. (R6). The bold numbers in the last column of Table 4.3 are the significant digits, whereas the italic numbers are the calculation data with errors. It might be that significant digits are lost in the addition and subtraction of several data in the recursion formula, especially for eccentricity functions whose absolute values are smaller and the significant digits are lost more seriously. For example, there are only five significant digits for $X_1^{-6,-5}(0.1)$. If the recursion of higher-order Hansen coefficients is required, this phenomenon may become more serious and require special attention.

4.5 Calculation of Hansen Coefficients for Large Orbital Eccentricities

The difficulty in calculating Hansen coefficients for large orbital eccentricities has received much attention. Taking an orbital eccentricity $e = 0.75$ as an example, the calculation of Hansen coefficients for large orbital eccentricities is discussed in the following.

Table 4.3 Simple example of recursion of Hansen coefficients $X_k^{-(n+1),m}(e)$ $(k = 1, e = 0.1)$

n	$-(n+1)$	m	Wnuk 's method	(R5) backward recursion	(R6) forward recursion
5	-6	-5	.220463393108E-10	.220463393108E-10	**.22046***5230850*E-10
5	-6	-4	.886030357479E-06	.886030357479E-06	
5	-6	-3	.709351254089E-04	.709351254089E-04	**.709351254089**E-04
5	-6	-2	.230562863411E-02	.230562863411E-02	
5	-6	-1	.379033046485E-01	.379033046485E-01	**.379033046485**E-01
5	-6	0	.315059918333E+00	.315059918333E+00	
5	-6	1	.106724073232E+01	.106724073232E+01	**.106724073232**E+01
5	-6	2	.105280679952E+00	.105280679952E+00	
5	-6	3	.393010452060E-02	.393010452060E-02	**.39301045206***1*E-02
5	-6	4	.438917338407E-04	.438917338407E-04	
5	-6	5	.262254655670E-06	.262254655670E-06	**.262254655***504*E-06
4	-5	-4	.263814353966E-08	.263814353966E-08	**.26381435***1998*E-08
4	-5	-3	.174907485179E-04	.174907485179E-04	
4	-5	-2	.105469787460E-02	.105469787460E-02	**.105469787460**E-02
4	-5	-1	.245399987149E-01	.245399987149E-01	
4	-5	0	.258630759867E+00	.258630759867E+00	**.258630759867**E+00
4	-5	1	.104103118694E+01	.104103118694E+01	
4	-5	2	.521120013220E-01	.521120013220E-01	**.521120013220**E-01
4	-5	3	.128624966754E-02	.128624966754E-02	
4	-5	4	-.209251647771E-04	-.209251647771E-04	-**.20925164777***7*E-04
3	-4	-3	.263043347113E-06	.263043347113E-06	.263043347113E-06
3	-4	-2	.341066515102E-03	.341066515102E-03	
3	-4	-1	.140614245716E-01	.140614245716E-01	.140614245716E-01
3	-4	0	.204322416608E+00	.204322416608E+00	
3	-4	1	.102037928865E+01	.102037928865E+01	.102037928865E+01
3	-4	2	.509311808276E-03	.509311808276E-03	
3	-4	3	.125210136568E-02	.125210136568E-02	.125210136568E-02
2	-3	-2	.209775890238E-04	.209775890238E-04	.209775890238E-04
2	-3	-1	.633435513970E-02	.633435513970E-02	
2	-3	0	.151708126135E+00	.151708126135E+00	.151708126135E+00
2	-3	1	.100508697100E+01	.100508697100E+01	
2	-3	2	-.499376309904E-01	-.499376309904E-01	-.499376309904E-01

4.5.1 Discrimination of the Stability of Direct Calculation Methods

As stated in Chap. 3, a stability criterion is required to compare the advantages and disadvantages of various direct methods of calculating Hansen coefficients. In fact, the stabilities of direct calculation methods can be determined by judging whether Hansen coefficients calculated with these methods meet the recursion relations of Hansen coefficients.

It is not difficult to verify that the calculation results of various direct calculation methods in Chap. 3 for small orbital eccentricities follow the recursion relations. However, for large orbital eccentricities, the calculation results may not follow the recursion relations. For example, the recursion formula of eccentricity function is [1]

$$G_{l-2,p-1,q} = \frac{l-2p}{l-2p+q}\sqrt{1-e^2}G_{lpq} + \frac{(l-1)e}{2(l-2p+q)\sqrt{1-e^2}}\left(G_{l-1,p-1,q-1} - G_{l-1,p,q+1}\right)$$

Table 4.4 presents the calculation results of Wnuk's method [5] and McClain's first method [2, 7], i.e., Eqs. (3.17) and (3.38), or Eqs. (3.87) and (3.76), obtained by the program computations for orbital eccentricity $e = 0.75$ and indexes $l = 30$, $p = 29$ and $q = -1$. To determine the stabilities of the calculation results of Wnuk's method and McClain's first method, we simply substitute the calculation results into the above recursion relation. For Wnuk's method, in this case, we have

$$G_{lpq} = 0.2056355791128 \times 10^3, \quad G_{l-1,p,q+1} = 0.4210025874464 \times 10^2$$

$$G_{l-1,p-1,q-1} = 0.9079971968072 \times 10^2, \quad G_{l-2,p-1,q} = 0.1542231019741 \times 10^2$$

Table 4.4 Data for stability discrimination (double precision computation)

l	p	q	k	Wnuk's method Eq. (3.17) or Eq. (3.87)[a]	McClain's first method Eq. (3.38) or Eq. (3.76)[b]
30	29	−1	−29	0.2056355791128E+03	−0.2798599055085E+09
29	29	0	−29	0.4210025874464E+02	−0.8434735464470E+09
29	28	−2	−29	0.9079971968072E+02	−0.2036490638135E+09
28	28	−1	−29	0.1542231019741E+02	−0.1645853936500E+09

[a] The summation ranges of the indexes s and t in Eq. (3.17) or Eq. (3.87) are 0 – 50 and −50 to 50, respectively.

[b] The summation range of the index i in Eq. (3.38) or Eq. (3.76) is 0 – 70.

Table 4.5 Data for stability discrimination

l	p	q	k	Wnuk's method Eq. (3.17) or Eq. (3.87)[a]	McClain's first method Eq. (3.38) or Eq. (3.76)[b]
30	29	−1	−29	0.6535635128884E+02	0.6539885877329E+02
29	29	0	−29	0.6202499298121E+00	0.6252491546172E+00
29	28	−2	−29	0.6271316802174E+02	0.6275657198574E+02
28	28	−1	−29	0.6538245252618E+01	0.6540499734199E+01

[a] Double-precision computation, where the summation ranges of the indexes s and t in Eq. (3.17) or Eq. (3.87) are $0 \sim 70$ and -70 to 70, respectively.
[b] Quadruple-precision computation, where the summation range of the index i in Eq. (3.38) or Eq. (3.76) is $0 \sim 70$.

Meanwhile, $l = 30$, $l - 2p = -28$, $l - 2p + q = -29$, and $l - 1 = 29$, then the right-hand side of the recursion formula $= 94.51164097 \neq 15.42231020$ (the left-hand side of the recursion formula). The result is thus unstable. Through similar discrimination, the calculation result of McClain's first method is even more unstable.

Further analysis reveals that the instability of the calculation result of Wnuk's method is due to the insufficient summation range, whereas the instability of the calculation result of McClain's first method is due to the insufficient word length of the computer. We expand the summation range for Wnuk's method, adopt the quadruple-precision computation for McClain's first method, and obtain the corresponding calculation results (see Table 4.5). Table 4.5 shows that the calculation results can now be considered stable. It is noted that the computational efficiency of McClain's first method is so high that it meets the requirements of practical application even with quadruple-precision computation.

4.5.2 Key Hansen Coefficients

Examining Hansen coefficients $X_{l-2p+q}^{-(l+1),l-2p} = G_{lpq}$ with the same index l, we find that for all l, the Hansen coefficients with the largest magnitudes occur at or near $p = l/2$. The perturbations caused by these functions are also the largest, and we should focus on these Hansen coefficients. For the convenience of expression, we refer to these Hansen coefficients with the largest magnitudes as key Hansen coefficients, and their corresponding indexes p are called key p.

Given the index l and examining the different indexes p, there are also Hansen coefficients with the largest magnitudes, and their corresponding indexes q are called key q. For different values of p, the values of key q change. As p decreases, the values of key q increase; as p increases, the values of key q decrease. When l is even, the key $p = l/2$, and the key $q = 0$ for the key p. When l is odd, there are two key p, namely $p = (l - 1)/2$ and $p = (l + 1)/2$, and the corresponding key

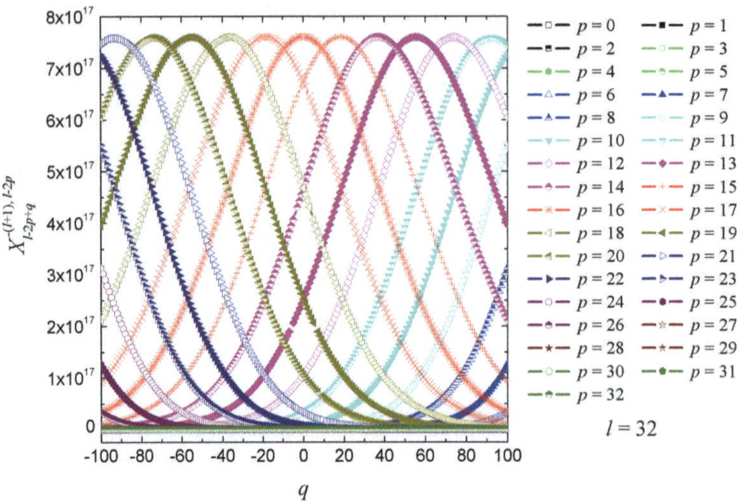

Fig. 4.1 Hansen coefficients $X_{l-2p+q}^{-(l+1),l-2p}$ versus p and q ($l = 32$, $e = 0.75$)

q for these two key p are opposite numbers, which is the reflection of the symmetry of eccentricity functions $G_{lpq} = G_{l,l-p,-q}$. Deviating from the key p, the values of key Hansen coefficients decrease and the absolute values of the corresponding $k = l - 2p + q$ increase. Their effect on perturbation thus decreases. This change is very regular. In the case that $l = 32$, the changes in Hansen coefficients and their derivatives for different p and q are presented in Figs. 4.1 and 4.2. This change regularity is helpful for us to choose the summation ranges of the indexes p and q. Accordingly, the summation range of the index q can be chosen, and the summation range of the index p can even be reduced.

It is noted that when the index l is even, the key $p = l/2$, and the key $q = 0$ for the key p. Then, $k = l - 2p + q = 0$, and the corresponding perturbation terms arising from tesseral harmonics of the Earth's gravitational field are the m-daily terms, which should be taken into account in the perturbation calculation generally. In this case, the involved Hansen coefficients are $X_0^{-(l+1),0}$. The calculation of Hansen coefficients $X_0^{-(n+1),m}$ is elaborated in detail in Chap. 2. For the short-period perturbation terms arising from tesseral harmonics of the Earth's gravitational field, Hansen coefficients $X_{l-2p+q}^{-(l+1),l-2p}$ with $k = l - 2p + q \neq 0$ are involved. In this case, when the index l is even, the key $p = l/2$, and the key q are $q = \pm 1$ for the key p.

In the case that the orbital eccentricity $e = 0.75$, the key Hansen coefficients and their derivatives with the index $l = 2, 3, \cdots, 32$ for the short-period perturbation terms arising from tesseral harmonics of the Earth's gravitational field ($m \neq 0$, $k = l - 2p + q \neq 0$) are listed in Table 4.6, together with the values of a^{l-1} for the orbital semi-major axis $a = 4.125$ (where the calculation unit of the orbital semi-major axis a is the Earth's equatorial radius). It is seen from Table 4.6 that the values of a^{l-1}, the key Hansen coefficients and their derivatives increase almost

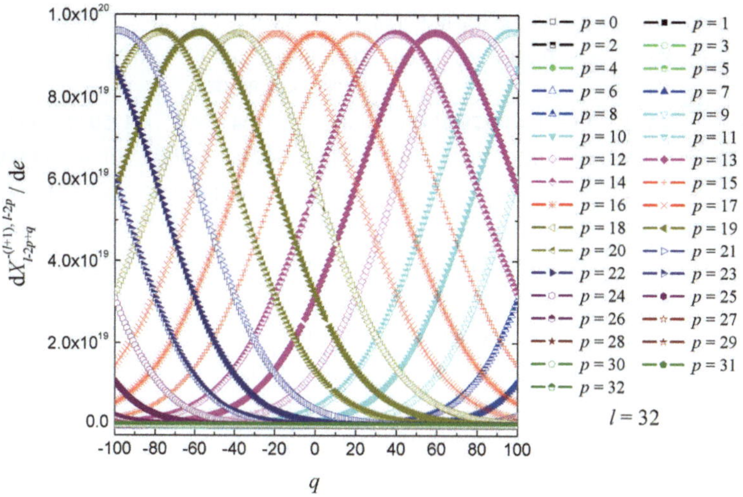

Fig. 4.2 Derivatives of Hansen coefficients $\frac{dX_{l-2p+q}^{-(l+1),l-2p}}{de}$ versus p and q ($l = 32$, $e = 0.75$)

synchronously, which ensures the convergence of the perturbation function expansion and perturbation calculation without the influence of the increase of the values of Hansen coefficients and their derivatives.

To calculate the short-period perturbation terms arising from tesseral harmonics of the Earth's gravitational field, we must not only calculate the Hansen coefficients and their derivatives but also perform summation operations. Figs. 4.1 and 4.2 show that for each value of p, the values of key Hansen coefficients and their derivatives are relatively large, and the peak values of the curves for different values of p are almost the same, which is like the need to calculate the Hansen coefficients and their derivatives with $q = -100$ to 100. This is difficult, and we need to solve this problem.

4.5.3 *Selection of the Summation Ranges*

(1) **Selection of the upper limit of the summation of the index l**

The disturbing function of the Earth's gravitational field is [8, 9]

$$R = \sum_{lmpq} \frac{\mu R_E^l}{a^{l+1}} J_{lm} \overline{F}_{lmp}(I) X_{l-2p+q}^{-(l+1),l-2p}(e) \cos(\psi + \lambda_{lm})$$

where l, m, p and q are the four indexes involved in the expansion of the disturbing function of the Earth's gravitational field, μ is the geocentric gravitational constant,

Table 4.6 Key Hansen coefficients, their derivatives and values of a^{l-1} ($e = 0.75, a = 4.125$)

l	Key p	Key q	Key Hansen coefficient	Derivative of key Hansen coefficient	Value of a^{l-1}
2	1	−1, 1	3.123424368790E+00	1.833211382647E+01	4.125000000000E+00
3	1, 2	5, −5	9.417387237031E+00	8.545437707750E+01	1.701562500000E+01
4	2	−1, 1	3.292885419256E+01	4.400631715619E+02	7.018945312500E+01
5	2, 3	7, −7	1.139204437566E+02	1.976825510046E+03	2.895314941406E+02
6	3	−1, 1	4.141483182415E+02	8.831778160292E+03	1.194317413330E+03
7	3, 4	8, −8	1.512471854307E+03	3.840249108203E+04	4.926559329987E+03
8	4	−1, 1	5.630365498401E+03	1.651051383419E+05	2.032205723619E+04
9	4, 5	8, −8	2.109134985382E+04	7.019592950119E+05	8.382848609930E+04
10	5	−1, 1	7.968077312176E+04	2.974111825335E+06	3.457925051596E+05
11	5, 6	8, −8	3.026739499739E+05	1.248508742408E+07	1.426394083783E+06
12	6	−1, 1	1.155330176303E+06	5.236707749868E+07	5.883875595607E+06
13	6, 7	9, −9	4.429024345322E+06	2.186181084491E+08	2.427098683188E+07
14	7	−1, 1	1.702593468743E+07	9.079490911798E+08	1.001178206815E+08
15	7, 8	9, −9	6.568977379414E+07	3.766672603201E+09	4.129860103112E+08
16	8	−1, 1	2.538449626584E+08	1.556783605810E+10	1.703567292534E+09
17	8, 9	9, −9	9.840000481751E+08	6.428414343114E+10	7.027215081701E+09
18	9	−1, 1	3.817903161528E+09	2.646902309889E+11	2.898726221202E+10
19	9, 10	9, −9	1.485299069762E+10	1.089069003564E+12	1.195724566246E+11
20	10	−1, 1	5.781515844350E+10	4.470796993939E+12	4.932363835763E+11
21	10, 11	9, −9	2.255629452175E+11	1.834269288367E+13	2.034600082252E+12
22	11	−1, 1	8.803001595765E+11	7.511559506084E+13	8.392725339291E+12
23	11, 12	9, −9	3.442401423670E+12	3.074668289544E+14	3.461999202458E+13
24	12	−1, 1	1.346366975804E+13	1.256562159377E+15	1.428074671014E+14
25	12, 13	9, −9	5.275054021131E+13	5.133491498735E+15	5.890808017932E+14
26	13	−1, 1	2.066891594816E+14	2.094383502820E+16	2.429958307397E+15
27	13, 14	9, −9	8.111158695398E+14	8.542330468765E+16	1.002357801801E+16
28	14	−1, 1	3.183064522796E+15	3.480055227070E+17	4.134725932430E+16
29	14, 15	9, −9	1.250864379851E+16	1.417423913789E+18	1.705574447127E+17
30	15	−1, 1	4.915316675233E+16	5.767167517751E+18	7.035494594400E+17
31	15, 16	9, −9	1.933902711374E+17	2.346124862324E+19	2.902141520190E+18
32	16	−1, 1	7.608158683394E+17	9.535359536375E+19	1.197133377078E+19

R_E is the mean equatorial radius of the Earth, $J_{lm} = \left(\overline{C}_{lm}^2 + \overline{S}_{lm}^2\right)^{1/2}$, $\overline{C}_{lm}$ and $\overline{S}_{lm}$ are the normalized coefficients of the Earth's gravitational field, $\psi = (l-2p)\omega + (l-2p+q)M + m(\Omega - \theta)$, a, e, I, Ω, ω and M are the Keplerian elements of the satellite, θ is the Greenwich sidereal time, $\lambda_{lm} = (l-m)\pi/2 - \arctan\left(\overline{S}_{lm}/\overline{C}_{lm}\right)$,

$\overline{F}_{lmp}(I)$ is the normalized inclination function [8], and $X_{l-2p+q}^{-(l+1),l-2p}(e)$ is the Hansen coefficient. Hansen coefficients $X_{l-2p+q}^{-(l+1),l-2p}(e)$ are often denoted as G_{lpq}, and they are referred to eccentricity functions [10].

Substituting the disturbing function of the Earth's gravitational field into the perturbation equations of motion and performing an integral calculation reveal that the perturbation solutions of satellite orbital elements a, e, I and Ω contain Hansen coefficients or eccentricity functions G_{lpq} but do not contain the derivatives of Hansen coefficients or the derivatives of eccentricity functions G'_{lpq}. In contrast, the perturbation solutions of two orbital elements ω and M contain the derivatives of Hansen coefficients or the derivatives of eccentricity functions G'_{lpq} [9–12]. In the following, we judge and select the relatively appropriate upper limit of summation of the index l by simply analyzing the maximum terms of the spatial position change of the satellite due to the short-period perturbation terms of tesseral harmonics of the Earth's gravitational field. Through deduction, the terms with the largest magnitudes in the perturbation solutions of the short-period perturbation terms arising from tesseral harmonics of the Earth's gravitational field are

$$\Delta\omega = \frac{\sqrt{1-e^2}}{e}\sum_{lmpq}\frac{R_E^l}{a^l}J_{lm}\overline{F}_{lmp}(I)\frac{G'_{lpq}}{l-2p+q}\sin(\psi+\lambda_{lm}) \tag{4.55}$$

$$\Delta M = -\frac{1-e^2}{e}\sum_{lmpq}\frac{R_E^l}{a^l}J_{lm}\overline{F}_{lmp}(I)\frac{G'_{lpq}}{l-2p+q}\sin(\psi+\lambda_{lm}) \tag{4.56}$$

The spatial position change of the satellite $\Delta\vec{r}$ is [9]

$$\Delta\vec{r} = \frac{\dot{\vec{r}}}{n}\Delta M + \left(\vec{R}\times\vec{r}\right)\Delta\omega = \left[\left(\vec{R}\times\vec{r}\right) - \sqrt{1-e^2}\frac{\dot{\vec{r}}}{n}\right]\Delta\omega \tag{4.57}$$

where $\vec{r}$ and $\dot{\vec{r}}$ are the position and velocity vectors of the satellite, respectively, n is the mean angular speed, and $\vec{R}$ is the unit vector normal to the orbital plane.

The projection of $\Delta\vec{r}$ in the direction $\frac{\vec{r}}{r}$ (radial, $r = |\vec{r}|$) is

$$\begin{aligned}\Delta r &= \frac{ae}{\sqrt{1-e^2}}\sin f\,\Delta M\\ &= -\sqrt{1-e^2}\sum_{lmpq}\frac{R_E^l}{a^{l-1}}J_{lm}\overline{F}_{lmp}(I)\frac{G'_{lpq}}{l-2p+q}\sin f\sin(\psi+\lambda_{lm})\end{aligned} \tag{4.58}$$

The projection of $\Delta\vec{r}$ in the direction $\vec{T}$ (transverse) is

$$
\begin{aligned}
\Delta\vec{r}\cdot\vec{T} &= r\left[1-\frac{a^2\left(1-e^2\right)}{r^2}\right]\Delta\omega \approx -2ae\cos f\,\Delta\omega \\
&= -2\sqrt{1-e^2}\sum_{lmpq}\frac{R_E^l}{a^{l-1}}J_{lm}\overline{F}_{lmp}(I)\frac{G'_{lpq}}{l-2p+q}\cos f\sin(\psi+\lambda_{lm})
\end{aligned}
\tag{4.59}
$$

In a comparison of these two terms, the magnitude of the along-track term (transverse) is larger.

Assuming the inclination functions have a value of 1, the orbital eccentricity $e = 0.75$, the orbital semi-major axis $a = 4.125$, and $J_{lm} = 10^{-6}/l$ and using the values of Hansen coefficients (or eccentricity functions), their derivatives and a^{l-1} in Table 4.6, it is appropriate to choose the upper limit of summation N of the index l as $N = 32$ to retain the short-period perturbation terms arising from tesseral harmonics of the Earth's gravitational field with magnitudes larger than 1 meter, according to Eq. (4.59). For a smaller upper limit of summation N of the index l, the perturbation calculation accuracy cannot be guaranteed to be 1 meter. If we choose $N = 18$, the perturbation calculation accuracy is approximately 2 meters.

It is noted that in actual calculations, the inclination functions need to be strictly calculated for each J_{lm} (using actual values) to make the judgement.

(2) **Selection of the summation ranges of the indexes p and q**

The magnitudes of the short-period perturbation terms arising from tesseral harmonics of the Earth's gravitational field ($m \neq 0$, $k = l - 2p + q \neq 0$) are proportional to $\left|G'_{lpq}\big/k\right|$. The change in $\left|G'_{lpq}\big/k\right|$ with the index q ($l = 32$) is presented in Fig. 4.3 and is completely different from the change in the derivatives of Hansen coefficients (see Fig. 4.2). Therefore, in the calculation of the short-period perturbation terms $\Delta\sigma$ arising from tesseral harmonics of the Earth's gravitational field, the summation ranges of the indexes p and q are chosen as

$$
\Delta\sigma = \sum_{l=2}^{32}\sum_{p=0}^{l}\sum_{q=-20}^{20}\Delta\sigma_{lpq}
\tag{4.60}
$$

As the perturbations are large near the key p, Eq. (4.60) is rewritten as

$$
\Delta\sigma = \sum_{l=2}^{32}\ \sum_{p=\text{key }p-\Delta p}^{\text{key }p+\Delta p}\ \sum_{q=-20}^{20}\Delta\sigma_{lpq}
\tag{4.61}
$$

Δp can be taken as 3 or 4 through the analysis. Of course, when $l = 2, 3$, the summation range of the index p remains $0 \sim l$.

It is noted that the above summation ranges are only the analytical results of key Hansen coefficients. The authors have no specific practice and hope that readers make further improvements in use.

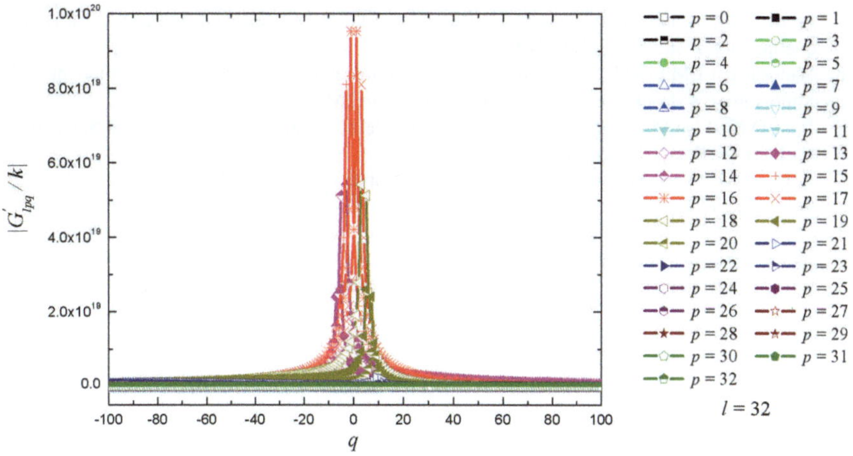

Fig. 4.3 $\left|G'_{lpq}\middle/k\right|$ versus the index q ($l = 32$, $p = 0 \sim 32$, $e = 0.75$, $a = 4.125$)

4.5.4 Computational Stability of Key Hansen Coefficients

To calculate the perturbations for orbits with large eccentricities, the summation ranges are determined, and the final question is whether Hansen coefficients can be accurately calculated.

The magnitudes of key Hansen coefficients are small for the index $l \leq 32$, and hence the loss of significant digits is small. The calculation results of key Hansen coefficients obtained with double-precision computation are stable. Moreover, in this case, the various methods of calculating the Hansen coefficients are all stable. Therefore, the calculation of Hansen coefficients related to the perturbations for orbits with large eccentricities is basically solved.

References

1. Giacaglia GEO (1976) A note on Hansen's coefficients in satellite theory. Celest Mech 14(4):515–523
2. McClain WD (1978) A recursively formulated first-order semianalytic artificial satellite theory based on the generalized method of averaging. Volume II. The explicit development of the first-order averaged equations of motion for the nonspherical gravitational and nonresonant third-body perturbations. NASA CR-156783
3. Vakhidov AA (2001) Some recurrence relations between Hansen coefficients. Celest Mech Dyn Astron 81(3):177–190
4. Wu LD, Zhang MJ (2021) Two new recursion formulae of Hansen coefficients. Chin Astron Astrophy 45(4):531–541
5. Wnuk E (1997) Highly eccentric satellite orbits. Adv Space Res 19(11):1735–1740
6. Wu LD, Zhang MJ (2022) Computational efficiency of the recursion of Hansen coefficients. Chin Astron Astrophy 46(2):65–72

7. Proulx RJ, McClain WD (1988) Series representations and rational approximations for Hansen coefficients. J Guidance 11(4):313–319
8. Gaposchkin EM (1973) 1973 Smithsonian standard earth (III). SAO special report no. 353, Smithsonian Astrophysical Observatory, Cambridge, Massachusetts
9. Wu LD (2011) Orbit and detection of artificial satellites and space debris. China Science and Technology Press, Beijing (In Chinese)
10. Kaula WM (1966) Theory of satellite geodesy: applications of satellites to geodesy. Blaisdell Publishing Company
11. Liu L, Tang JS (2015) Orbit theory of satellites and application. Publishing House of Electronics Industry, Beijing (In Chinese)
12. Liu L (2000) Orbit theory of spacecraft. National Defense Industry Press, Beijing (In Chinese)

Chapter 5
Other Methods of Calculating Hansen Coefficients

5.1 Harmonic Analysis Method

Sharaf and Selim proposed the harmonic analysis method of calculating Hansen coefficients [1]. In the following, the principle of harmonic analysis is first introduced.

Suppose $f(x)$ is a function on the interval $[a, b]$. It is approximately expressed as [2]

$$f(x) = \frac{1}{2}A_0 + \sum_{j=1}^{N}\left[A_j \cos\left(\frac{2\pi jx}{b-a}\right) + B_j \sin\left(\frac{2\pi jx}{b-a}\right)\right] \tag{5.1}$$

where $A_j\ (j = 0, 1, \ldots, N)$ and $B_j\ (j = 1, 2, \ldots, N)$ are Fourier coefficients, which can be solved using $2N + 1$ function values $f_n = f(x_n)$, $x_n = (b - a)n/(2N + 1)$ and $n = 0, 1, \ldots, 2N$.

It is widely known that the Fourier coefficients A_j and B_j can be calculated according to

$$A_j = \frac{2}{2N+1}\sum_{n=0}^{2N} f_n \cos\left(\frac{2\pi nj}{2N+1}\right)$$

$$B_j = \frac{2}{2N+1}\sum_{n=0}^{2N} f_n \sin\left(\frac{2\pi nj}{2N+1}\right) \tag{5.2}$$

Moreover, there is a method of calculating A_j and B_j more efficiently. Using the relations

$$U_{2N+2,j} = U_{2N+1,j} = 0$$

$$U_{n,j} = f_n + 2\cos\left(\frac{2\pi j}{2N+1}\right)U_{n+1,j} - U_{n+2,j}, \quad (n = 2N, 2N-1, \cdots, 1) \tag{5.3}$$

L.D Wu and M.J Zhang, *Hansen Coefficients in Satellite Orbital Dynamics*,
Springer Aerospace Technology, https://doi.org/10.1007/978-981-97-0456-9_5

A_j and B_j are calculated according to

$$\begin{aligned} A_j &= \frac{2}{2N+1}\left[f_0 + U_{1,j}\cos\left(\frac{2\pi j}{2N+1}\right) - U_{2,j}\right] \\ B_j &= \frac{2}{2N+1}U_{1,j}\sin\left(\frac{2\pi j}{2N+1}\right) \end{aligned} \tag{5.4}$$

Once A_j and B_j are derived, the trigonometric polynomial of the function $f(x)$ is obtained from Eq. (5.1); this is an approximation of $f(x)$.

The harmonic analysis method of calculating Hansen coefficients is as follows. Through the harmonic analysis of the functions $(r/a)^n \cos mf$ and $(r/a)^n \sin mf$, the Fourier coefficients $A_j^{(c)}$, $B_j^{(c)}$, $A_j^{(s)}$ and $B_j^{(s)}$ are obtained. According to the odd and even properties of the functions $(r/a)^n \cos mf$ and $(r/a)^n \sin mf$, $B_j^{(c)} = 0$, $A_j^{(s)} = 0$, we have

$$\left(\frac{r}{a}\right)^n \cos mf = \frac{1}{2}A_0^{(c)} + \sum_{j=1}^{\infty} A_j^{(c)} \cos jM \tag{5.5}$$

$$\left(\frac{r}{a}\right)^n \sin mf = \sum_{j=1}^{\infty} B_j^{(s)} \sin jM \tag{5.6}$$

Then, using Eqs. (1.9)–(1.11), the Hansen coefficients are obtained as

$$\begin{aligned} X_0^{n,m} &= \frac{A_0^{(c)}}{2} \\ X_j^{n,m} &= \frac{A_j^{(c)} + B_j^{(s)}}{2} \\ X_{-j}^{n,m} &= \frac{A_j^{(c)} - B_j^{(s)}}{2}, \quad (j = 1, 2, \cdots, N) \end{aligned} \tag{5.7}$$

The principle of this method is simple. The calculation of functions $(r/a)^n \cos mf$ and $(r/a)^n \sin mf$ in terms of the mean anomaly M is straightforward, and the calculation quantity is small. The method is thus effective. The order N of harmonic analysis needs to be set in this method. Sharaf and Selim [1] chose an appropriate N through testing. N relates to the value of the eccentricity e and the index j of Hansen coefficients $X_j^{n,m}$ to be calculated. Generally, a larger e corresponds to larger N. For $e \le 0.1$, the order $N = 20$ can be chosen, whereas for large eccentricities, N may need to be taken as 100. A calculation example of Hansen coefficients for large eccentricity using the harmonic analysis method is presented in Table 5.1.

Table 5.1 Calculation example of Hansen coefficients using the harmonic analysis method ($e = 0.786, n = 8, m = 4$)

n	m	j	$A_j^{(c)}$	$B_j^{(s)}$	$X_j^{n,m}$	$X_{-j}^{n,m}$
8	4	0	56.8135826252905	0.0000000000000	28.4067913126453	28.4067913126453
8	4	1	−47.0631379295400	−25.6930380703344	−36.3780879999372	−10.6850499296028
8	4	2	23.9162065946344	21.1464080624723	22.5313073285533	1.3848992660810
8	4	3	−4.7040536019876	−4.8420348582502	−4.7730442301189	0.0689906281313
8	4	4	−0.6054642529848	−0.6192405815557	−0.6123524172703	0.0068881642855
8	4	5	−0.0262285201957	−0.0283248489909	−0.0272766845933	0.0010481643976
8	4	6	0.0293642636330	0.0289444711152	0.0291543673741	0.0002098962589
8	4	7	0.0217702734691	0.0216686121556	0.0217194428124	0.0000508306568
8	4	8	0.0121886747256	0.0121604544194	0.0121745645725	0.0000141101531
8	4	9	0.0063739605771	0.0063653088749	0.0063696347260	0.0000043258511
8	4	10	0.0032595654106	0.0032567184673	0.0032581419390	0.0000014234716
8	4	11	0.0016462169684	0.0016452362422	0.0016457266053	0.0000004903631
8	4	12	0.0008173967224	0.0008170522339	0.0008172244782	0.0000001722442
8	4	13	0.0003926830122	0.0003925639804	0.0003926234963	0.0000000595159
8	4	14	0.0001761427525	0.0001761049769	0.0001761238647	0.0000000188878
8	4	15	0.0000672269906	0.0000672181321	0.0000672225613	0.0000000044292
8	4	16	0.0000140574113	0.0000140581826	0.0000140577969	−0.0000000003856
8	4	17	−0.0000103329992	−0.0000103296388	−0.0000103313190	−0.0000000016802
8	4	18	−0.0000200483751	−0.0000200448646	−0.0000200466198	−0.0000000017553
8	4	19	−0.0000224846486	−0.0000224817097	−0.0000224831792	−0.0000000014695
8	4	20	−0.0000215082507	−0.0000215059905	−0.0000215071206	−0.0000000011301

(continued)

Table 5.1 (continued)

n	m	j	$A_j^{(c)}$	$B_j^{(s)}$	$X_j^{n,m}$	$X_{-j}^{n,m}$
8	4	21	−0.0000191169486	−0.0000191152787	−0.0000191161136	−0.0000000008350
8	4	22	−0.0000163168462	−0.0000163156375	−0.0000163162419	−0.0000000006044
8	4	23	−0.0000135894593	−0.0000135885933	−0.0000135890263	−0.0000000004330
8	4	24	−0.0000111418821	−0.0000111412646	−0.0000111415734	−0.0000000003087
8	4	25	−0.0000090410257	−0.0000090405860	−0.0000090408059	−0.0000000002199
8	4	26	−0.0000072854084	−0.0000072850950	−0.0000072852517	−0.0000000001567
8	4	27	−0.0000058431422	−0.0000058429185	−0.0000058430303	−0.0000000001119
8	4	28	−0.0000046715968	−0.0000046714367	−0.0000046715168	−0.0000000000800
8	4	29	−0.0000037271649	−0.0000037270498	−0.0000037271074	−0.0000000000576
8	4	30	−0.0000029697338	−0.0000029696511	−0.0000029696925	−0.0000000000413
8	4	31	−0.0000023643859	−0.0000023643262	−0.0000023643561	−0.0000000000299
8	4	32	−0.0000018816956	−0.0000018816526	−0.0000018816741	−0.0000000000215
8	4	33	−0.0000014973712	−0.0000014973400	−0.0000014973556	−0.0000000000156
8	4	34	−0.0000011916275	−0.0000011916051	−0.0000011916163	−0.0000000000112
8	4	35	−0.0000009484988	−0.0000009484828	−0.0000009484908	−0.0000000000080
8	4	36	−0.0000007551808	−0.0000007551694	−0.0000007551751	−0.0000000000057
8	4	37	−0.0000006014489	−0.0000006014411	−0.0000006014450	−0.0000000000039
8	4	38	−0.0000004791622	−0.0000004791571	−0.0000004791596	−0.0000000000025
8	4	39	−0.0000003818505	−0.0000003818476	−0.0000003818490	−0.0000000000015
8	4	40	−0.0000003043780	−0.0000003043766	−0.0000003043773	−0.0000000000007
8	4	41	−0.0000002426696	−0.0000002426696	−0.0000002426696	0.0000000000000

(continued)

Table 5.1 (continued)

n	m	j	$A_j^{(c)}$	$B_j^{(s)}$	$X_j^{n,m}$	$X_{-j}^{n,m}$
8	4	42	−0.0000001934933	−0.0000001934945	−0.0000001934939	0.0000000000006
8	4	43	−0.0000001542851	−0.0000001542873	−0.0000001542862	0.0000000000011
8	4	44	−0.0000001230104	−0.0000001230135	−0.0000001230120	0.0000000000016
8	4	45	−0.0000000980539	−0.0000000980577	−0.0000000980558	0.0000000000019
8	4	46	−0.0000000781320	−0.0000000781369	−0.0000000781345	0.0000000000024
8	4	47	−0.0000000622252	−0.0000000622310	−0.0000000622281	0.0000000000029
8	4	48	−0.0000000495217	−0.0000000495282	−0.0000000495250	0.0000000000033
8	4	49	−0.0000000393755	−0.0000000393830	−0.0000000393793	0.0000000000038
8	4	50	−0.0000000312718	−0.0000000312806	−0.0000000312762	0.0000000000044
8	4	51	−0.0000000248007	−0.0000000248104	−0.0000000248056	0.0000000000049
8	4	52	−0.0000000196343	−0.0000000196455	−0.0000000196399	0.0000000000056
8	4	53	−0.0000000155115	−0.0000000155240	−0.0000000155178	0.0000000000063
8	4	54	−0.0000000122231	−0.0000000122373	−0.0000000122302	0.0000000000071
8	4	55	−0.0000000096028	−0.0000000096186	−0.0000000096107	0.0000000000079
8	4	56	−0.0000000075162	−0.0000000075340	−0.0000000075251	0.0000000000089
8	4	57	−0.0000000058567	−0.0000000058769	−0.0000000058668	0.0000000000101
8	4	58	−0.0000000045389	−0.0000000045617	−0.0000000045503	0.0000000000114
8	4	59	−0.0000000034945	−0.0000000035199	−0.0000000035072	0.0000000000127
8	4	60	−0.0000000026677	−0.0000000026963	−0.0000000026820	0.0000000000143
8	4	61	−0.0000000020155	−0.0000000020476	−0.0000000020316	0.0000000000160
8	4	62	−0.0000000015023	−0.0000000015379	−0.0000000015201	0.0000000000178

(continued)

Table 5.1 (continued)

n	m	j	$A_j^{(c)}$	$B_j^{(s)}$	$X_j^{n,m}$	$X_{-j}^{n,m}$
8	4	63	−0.0000000010991	−0.0000000011393	−0.0000000011192	0.0000000000201
8	4	64	−0.0000000007845	−0.0000000008293	−0.0000000008069	0.0000000000224
8	4	65	−0.0000000005397	−0.0000000005894	−0.0000000005645	0.0000000000248
8	4	66	−0.0000000003498	−0.0000000004056	−0.0000000003777	0.0000000000279
8	4	67	−0.0000000002041	−0.0000000002660	−0.0000000002350	0.0000000000310
8	4	68	−0.0000000000925	−0.0000000001615	−0.0000000001270	0.0000000000345
8	4	69	−0.0000000000081	−0.0000000000846	−0.0000000000463	0.0000000000382
8	4	70	0.0000000000556	−0.0000000000294	0.0000000000131	0.0000000000425
8	4	71	0.0000000001027	0.0000000000085	0.0000000000556	0.0000000000471
8	4	72	0.0000000001372	0.0000000000333	0.0000000000853	0.0000000000520
8	4	73	0.0000000001625	0.0000000000478	0.0000000001051	0.0000000000574
8	4	74	0.0000000001803	0.0000000000539	0.0000000001171	0.0000000000632
8	4	75	0.0000000001926	0.0000000000541	0.0000000001234	0.0000000000692
8	4	76	0.0000000002008	0.0000000000498	0.0000000001253	0.0000000000755
8	4	77	0.0000000002068	0.0000000000419	0.0000000001243	0.0000000000824
8	4	78	0.0000000002100	0.0000000000315	0.0000000001208	0.0000000000892
8	4	79	0.0000000002123	0.0000000000195	0.0000000001159	0.0000000000964
8	4	80	0.0000000002131	0.0000000000066	0.0000000001099	0.0000000001033

For the calculation example in Table 5.1, the order N is set at 80. The partial Hansen coefficients derived using the harmonic analysis method and the direct calculation method are compared in Table 5.2. The table shows that the differences between the calculation results of these two methods are 10^{-13}, and the accuracy of the harmonic analysis method is thus good. Of course, smaller magnitudes of Hansen coefficients correspond to larger relative error. This is a common fault of purely mathematical methods, which has no impact on perturbation calculations.

However, there remain problems in the calculation of higher-order Hansen coefficients for large eccentricities when using the harmonic analysis method. The calculation results for the same example considered in Table 3.5 (but with a greatly expanded range of the index q) obtained using the harmonic analysis method, are presented in Table 5.3, together with the calculation results obtained using Eq. (3.17), for the comparison of calculation results. The table shows that when $-6 \leq q \leq 74$, the calculation results derived by the two methods are in relatively good agreement, whereas at both ends of the values of the index q, it is obvious that the calculation results derived by the harmonic analysis method seriously diverge and their magnitudes are much larger, which affects the results of the perturbation calculations. Of course, the index range in this calculation example exceeds the demands of perturbation calculations, but it shows that the harmonic analysis method also has a stability problem. Therefore, care must be taken in adopting this method.

The advantage of the harmonic analysis method is that it only calculates Hansen coefficients required in the perturbation calculations, and Hansen coefficients for all values of the index q can be calculated at one time. Its disadvantage is that it cannot calculate the kernels of Hansen coefficients required in non-singular perturbation theory. The harmonic analysis method of calculating the derivatives of Hansen coefficients needs further study.

5.2 Taylor Series Expansion Method

Any function $f(x)$ can be expanded as a Tayler series:

$$f(x) = \sum_{n=0}^{\infty} \frac{f^{(n)}(a)}{n!}(x-a)^n \tag{5.8}$$

where $f^{(n)}(\cdot)$ refers to the n-th order derivative of the function $f(\cdot)$. Hansen coefficients can be calculated using Tayler series expansion in the vicinity of the eccentricity e^*, and we have

$$X_k^{n,m}(e) = \sum_{l=0}^{\infty} \frac{(e-e^*)^l}{l!} \frac{d^l}{de^l} X_k^{n,m}(e)|_{e=e^*} \tag{5.9}$$

Table 5.2 Comparison of Hansen coefficients derived using the harmonic analysis method and the direct calculation method

n	m	j	Direct calculation method, Eq. (3.39)		Harmonic analysis method	
			$X_j^{n,m}$	$X_{-j}^{n,m}$	$X_j^{n,m}$	$X_{-j}^{n,m}$
8	4	0	28.4067913126452	28.4067913126452	28.4067913126453	28.4067913126453
8	4	1	−36.3780879999372	−10.6850499296028	−36.3780879999372	−10.6850499296028
8	4	2	22.5313073285533	1.3848992660810	22.5313073285533	1.3848992660810
8	4	3	−4.7730442301190	0.0689906281313	−4.7730442301189	0.0689906281313
8	4	4	−0.6123524172703	0.0068881642854	−0.6123524172703	0.0068881642855
8	4	5	−0.0272766845933	0.0010481643976	−0.0272766845933	0.0010481643976
8	4	6	0.0291543673741	0.0002098962589	0.0291543673741	0.0002098962589
8	4	7	0.0217194428124	0.0000508306567	0.0217194428124	0.0000508306568
8	4	8	0.0121745645725	0.0000141101531	0.0121745645725	0.0000141101531
8	4	9	0.0063696347260	0.0000043258511	0.0063696347260	0.0000043258511
8	4	10	0.0032581419390	0.0000014234716	0.0032581419390	0.0000014234716
8	4	11	0.0016457266053	0.0000004903630	0.0016457266053	0.0000004903631
8	4	12	0.0008172244782	0.0000001722442	0.0008172244782	0.0000001722442
8	4	13	0.0003926234963	0.0000000595158	0.0003926234963	0.0000000595159
8	4	14	0.0001761238647	0.0000000188878	0.0001761238647	0.0000000188878
8	4	15	0.0000672225613	0.0000000044292	0.0000672225613	0.0000000044292
8	4	16	0.0000140577970	−0.0000000003857	0.0000140577969	−0.0000000003856
8	4	17	−0.0000103313190	−0.0000000016803	−0.0000103313190	−0.0000000016802
8	4	18	−0.0000200466199	−0.0000000017554	−0.0000200466198	−0.0000000017553
8	4	19	−0.0000224831792	−0.0000000014696	−0.0000224831792	−0.0000000014695

(continued)

Table 5.2 (continued)

n	m	j	Direct calculation method, Eq. (3.39)		Harmonic analysis method	
			$X_j^{n,m}$	$X_{-j}^{n,m}$	$X_j^{n,m}$	$X_{-j}^{n,m}$
8	4	20	−0.0000215071206	−0.0000000011303	−0.0000215071206	−0.0000000011301
8	4	21	−0.0000191161136	−0.0000000008351	−0.0000191161136	−0.0000000008350
8	4	22	−0.0000163162419	−0.0000000006045	−0.0000163162419	−0.0000000006044
8	4	23	−0.0000135890263	−0.0000000004331	−0.0000135890263	−0.0000000004330
8	4	24	−0.0000111415733	−0.0000000003089	−0.0000111415734	−0.0000000003087
8	4	25	−0.0000090408059	−0.0000000002201	−0.0000090408059	−0.0000000002199
8	4	26	−0.0000072852517	−0.0000000001569	−0.0000072852517	−0.0000000001567
8	4	27	−0.0000058430303	−0.0000000001122	−0.0000058430303	−0.0000000001119
8	4	28	−0.0000046715168	−0.0000000000804	−0.0000046715168	−0.0000000000800

Table 5.3 Calculation example of higher-order Hansen coefficients for large eccentricity ($e = 0.6$)

l	p	q	Harmonic analysis method	Calculation with Eq. (3.17)
30	2	−186	2.268466167450E+02	3.130302325149E−47
30	2	−176	3.432091712952E+01	−9.402087632993E−43
30	2	−166	4.983093261719E+00	1.480147133286E−38
30	2	−156	6.966361999512E−01	−1.110866463879E−33
30	2	−146	9.328079223633E−02	2.265103007414E−28
30	2	−136	1.224517822266E−02	2.392533616964E−24
30	2	−126	1.613616943359E−03	5.451479761208E−21
30	2	−116	1.111626625061E−04	3.428238507259E−18
30	2	−106	7.376074790955E−05	6.247551121997E−16
30	2	−96	5.383789539337E−05	3.559020535456E−14
30	2	−86	5.576205148827E−04	8.984190535882E−13
30	2	−76	2.038657257799E−04	1.775501402688E−11
30	2	−66	6.383087089148E−04	3.456231620913E−10
30	2	−56	4.687666082059E−04	6.725445195169E−09
30	2	−46	2.682225233844E−04	1.308274274529E−07
30	2	−36	6.117355796485E−05	2.544081301439E−06
30	2	−26	−6.089675453515E−06	4.945521622123E−05
30	2	−16	9.015487705499E−04	9.610469516074E−04
30	2	−6	2.826079976261E−02	2.825215025931E−02
30	2	4	−7.368408341527E+00	−7.368422870064E+00
30	2	14	−8.444944467696E+02	−8.444945803511E+02
30	2	24	3.703967948391E+04	3.703967879355E+04
30	2	34	−3.111009206984E+05	−3.111009204075E+05
30	2	44	7.385191162567E+06	7.385191158502E+06
30	2	54	−1.657042173391E+08	−1.657042173349E+08
30	2	64	9.097157363313E+08	9.097157363731E+08
30	2	74	1.036022271288E+10	1.036019351604E+10
30	2	84	3.054074274299E+10	3.029021053709E+10
30	2	94	4.816560275550E+10	2.749158799765E+10
30	2	104	4.989883893148E+10	−5.315134646163E+09
30	2	114	3.772795010027E+10	2.592112239776E+08
30	2	124	2.223235091628E+10	−7.591437198446E+06
30	2	134	1.068327297603E+10	8.155769913265E+04
30	2	144	4.326959508444E+09	−1.188980024924E+02
30	2	154	1.514795256822E+09	−9.691678518646E−01
30	2	164	4.675082790336E+08	1.264692638151E−03
30	2	174	1.292301136263E+08	2.219407647324E−06

(continued)

Table 5.3 (continued)

l	p	q	Harmonic analysis method	Calculation with Eq. (3.17)
30	2	184	3.241098116409E+07	−8.633146180139E−10
30	2	194	7.454694926142E+06	−1.687662323724E−12

The key here is how to derive the derivatives $\frac{d^l}{de^l}X_k^{n,m}(e)$. de Moraes and Wnuk presented the following calculation method [3].

Giacaglia's expression of Hansen coefficients is [4]

$$X_k^{n,m} = \sum_{s=0}^{s_1}\sum_{j=-s}^{j_1}\binom{n-m+1}{s}\binom{n+m+1}{s+j}f_{n,s,j}(\beta)J_{k-m+j}(ke) \tag{5.10}$$

where

$$f_{n,s,j}(\beta) = \left(1+\beta^2\right)^{-n-1}(-\beta)^j\beta^{2s}$$

$$\beta = \frac{e}{1+\sqrt{1-e^2}}$$

$$s_1 = \begin{cases} n-m+1, & (n-m+1 \geq 0) \\ \infty, & (n-m+1 < 0) \end{cases}$$

$$j_1 = \begin{cases} n+m+1-s, & (n+m+1 \geq 0) \\ \infty, & (n+m+1 < 0) \end{cases}$$

Following Ref. [5], $f_{n,s,j}(\beta)$ is expressed as

$$\begin{aligned} &f_{n,s,j}(\beta) \\ &= \sum_{i=0}^{\infty}\left[2\binom{-n+2i-2+j+2s}{i} - \binom{-n+2i-1+j+2s}{i}\right] \\ &\times (-1)^j\left(\frac{e}{2}\right)^{2i+j+2s} \end{aligned} \tag{5.11}$$

Using Eq. (5.10), the derivatives are calculated according to

$$\begin{aligned} \frac{d^l}{de^l}X_k^{n,m}(e) = &\sum_{s=0}^{s_1}\sum_{j=-s}^{j_1}\binom{n-m+1}{s}\binom{n+m+1}{s+j} \\ &\times\left[\sum_{r=0}^{l}\binom{l}{r}\frac{d^r}{de^r}f_{n,s,j}(\beta)\frac{d^{l-r}}{de^{l-r}}J_{k-m+j}(ke)\right] \end{aligned} \tag{5.12}$$

in which the derivatives of $f_{n,s,j}(\beta)$ are obtained by directly differentiating Eq. (5.11) with respect to the eccentricity, and the derivatives of Bessel functions are expressed as

$$\frac{d^t}{de^t} J_l(x) = \frac{1}{2^t} \sum_{r=0}^{t} (-1)^r \binom{t}{r} J_{l-t+2r}(x) \tag{5.13}$$

Given e^* and the upper limits of the summation of Eqs. (5.9), (5.11) and (5.12), Hansen coefficients can be calculated from Eq. (5.9) immediately.

Another option in calculating the derivatives $\frac{d^l}{de^l} X_k^{n,m}(e)$ is to use Eq. (3.10); i.e.,

$$\begin{aligned} &X_{m+s}^{n,m} \\ &= \left(-\frac{e}{2}\right)^s \sum_{t=0}^{\infty} \left(\frac{e}{2}\right)^{2t} \left\{ \sum_{j=0}^{t} \sum_{p=0}^{j} \binom{n+m+1}{j-p} X_p \sum_{q=0}^{s+j} \binom{n-m+1}{s+j-q} X_q \left[\binom{k}{t-j} - \binom{k}{t-j-1} \right] \right\} \end{aligned} \tag{5.14}$$

By directly differentiating Eq. (5.14) with respect to the eccentricity, the derivatives $\frac{d^l}{de^l} X_k^{n,m}(e)$ are immediately obtained, but we note the sign of the index s.

However, the formulae seem complex. To calculate the derivatives using Eq. (5.14), a quintuple summation is required for the calculation of Hansen coefficients, and a sextuple summation is required for the calculation of Hansen coefficients using Eq. (5.12). de Moraes and Wnuk calculated the 70×70 order orbital perturbation of a satellite using Eq. (5.12), performed the tests for the satellite with a large orbital eccentricity $e = 0.723$, and obtained satisfactory results [3]. However, it is unclear whether instability occurs when Eq. (5.14) is adopted to calculate the derivatives.

Vakhidov adopted a Tayler series to calculate Hansen coefficients. The calculation formula is [6]

$$X_s^{n,m}(e) = \sum_{k=0}^{N} A_k (e - e_0)^k \tag{5.15}$$

Vakhidov first derived an expression for A_k through computer algebra and then calculated the values of A_k for each n, m, s and e_0. Vakhidov [6] found that the derivation of the coefficients A_k is complex for $e_0 \neq 0$ and requires a lot of computing resources, including much computer memory and computing time. The expansion coefficients of Eq. (5.9) can be easily obtained only in the case that $N \leq 15$. Vakhidov thus recommended using a Tayler series of $e_0 = 0$ to calculate Hansen coefficients of small eccentricities. Here, it needs to expand to $N = 20$ for $e < 0.1$ and to $N = 50$ for $e < 0.4$, and the corresponding calculation accuracy is approximately 12 significant digits [6].

Da Silva Fernandes played an important role in developing the above Taylor series expansion method of calculating Hansen coefficients. He proved that the convergence range of Eq. (5.9) is the same as that of Kepler's equation [7]. The specific convergence radii are given in Table 5.4. Of course, the convergence radii here refer to the convergence of the series of $N = \infty$, but a calculation up to $N = \infty$ is impossible. Therefore, the convergence radius is not equal to the available radius, and indeed, the available radius may be much smaller than the convergence radius. For example, when $e_0 = 0$ and $N = 50$, Vakhidov's calculation [6] is applicable only to the cases that $e < 0.4$, which is a value of e much smaller than 0.662743.

To calculate the derivatives of $\frac{d^l}{de^l} X_k^{n,m}(e)$ specifically, the expressions of $\frac{d^r}{de^r} f_{n,s,j}(\beta)$ and $\frac{d^{l-r}}{de^{l-r}} J_{k-m+j}(ke)$ in Eq. (5.12) are further considered. According to Eq. (5.13), the expression $\frac{d^{l-r}}{de^{l-r}} J_{k-m+j}(ke)$ can be written as

$$\frac{d^{l-r}}{de^{l-r}} J_{k-m+j}(ke) = \left(\frac{k}{2}\right)^{l-r} \sum_{i=0}^{l-r} (-1)^i \binom{l-r}{i} J_{k-m+j-l+r+2i}(ke) \tag{5.16}$$

Rewriting Eq. (5.11) as

$$f_{n,s,j}(\beta) = \sum_{i=0}^{\infty} A_{nsji} e^{2i+j+2s} \tag{5.17}$$

Table 5.4 Convergence radii of Eq. (5.9) [7]

e^*	Convergence radius
0	0.662743
0.05	0.658512
0.10	0.646480
0.15	0.628099
0.20	0.604830
0.25	0.577822
0.30	0.547916
0.35	0.515719
0.40	0.481673
0.45	0.446109
0.50	0.409278
0.55	0.371375
0.60	0.332554
0.65	0.292938
0.70	0.252628
0.75	0.211707
0.80	0.170243
0.85	0.128294
0.90	0.085910
0.95	0.043134
1.00	0

where

$$A_{nsji} = \frac{(-1)^j}{2^{2i+j+2s}} \left[2\binom{-n+2i-2+j+2s}{i} - \binom{-n+2i-1+j+2s}{i} \right] \tag{5.18}$$

$\frac{d^r}{de^r} f_{n,s,j}(\beta)$ is expressed as

$$\frac{d^r}{de^r} f_{n,s,j}(\beta) = \sum_{i=0}^{\infty} A_{nsji}(2i+j+2s-r+1)_r e^{2i+j+2s-r} \tag{5.19}$$

Similarly, if Eq. (3.10) is adopted to calculate the derivatives, we have

$$X_{m+s}^{n,m} = \sum_{t=0}^{\infty} A_{st} e^{s+2t} \tag{5.20}$$

where

$$A_{st} = \begin{cases} (-1)^s \frac{1}{2^{s+2t}} \left\{ \sum_{j=0}^{t} \sum_{p=0}^{j} \binom{n+m+1}{j-p} X_p \sum_{q=0}^{s+j} \binom{n-m+1}{s+j-q} X_q \left[\binom{k}{t-j} - \binom{k}{t-j-1} \right] \right\}, & (s \geq 0) \\ (-1)^{|s|} \frac{1}{2^{|s|+2t}} \left\{ \sum_{j=0}^{t} \sum_{p=0}^{j} \binom{n-m+1}{j-p} X_p \sum_{q=0}^{|s|+j} \binom{n+m+1}{|s|+j-q} X_q \left[\binom{k}{t-j} - \binom{k}{t-j-1} \right] \right\}, & (s < 0) \end{cases}$$

$$k = 2t - n + |s| - p - q - 2, \quad X_p = \begin{cases} \frac{(m+s)^p}{p!}, & (s \geq 0) \\ \frac{(-m+|s|)^p}{p!}, & (s < 0) \end{cases}$$

$$X_q = \begin{cases} \frac{(m+s)^q}{q!}(-1)^q, & (s \geq 0) \\ \frac{(-m+|s|)^q}{q!}(-1)^q, & (s < 0) \end{cases}$$

By differentiating Eq. (5.20) with respect to the eccentricity, we obtain

$$\frac{d^l}{de^l} X_{m+s}^{n,m} = \sum_{t=0}^{\infty} A_{st}(s+2t-l+1)_l e^{s+2t-l} \tag{5.21}$$

In particular, when $e^* = 0$, Eqs. (5.16), (5.19) and (5.21) become

$$\frac{d^{l-r} J_{k-m+j}(0)}{de^{l-r}}$$

$$= \begin{cases} 0, & (l - r < m - k - j) \\ \left(\frac{k}{2}\right)^{l-r} (-1)^{(l-r+m-k-j)/2} \binom{l-r}{\frac{l-r+m-k-j}{2}}, & (l - r \geq m - k - j;\ \text{even}) \\ 0, & (l - r \geq m - k - j;\ \text{odd}) \end{cases} \tag{5.22}$$

$$\frac{d^r}{de^r} f_{n,s,j}(\beta) = \begin{cases} 0, & (r < j + 2s) \\ A_{nsj,(r-j-2s)/2} r!, & (r \geq j + 2s;\ \text{even}) \\ 0, & (r \geq j + 2s;\ \text{odd}) \end{cases} \tag{5.23}$$

$$\frac{d^l}{de^l} X_{m+s}^{n,m} = \begin{cases} 0, & (l < s) \\ A_{stl} l!, & (l \geq s;\ \text{even}) \\ 0, & (l \geq s;\ \text{odd}) \end{cases} \tag{5.24}$$

Substituting Eq. (5.24) into Eq. (5.9), we obtain Eq. (5.20); i.e., Eq. (3.10). Therefore, using $e^* = 0$, we do not derive a new expression. Such a Tayler expansion of $e^* = 0$ is meaningless. In addition, it seems that by substituting Eqs. (5.22) and (5.23) into Eq. (5.12), and then substituting into Eq. (5.9), there is the suspicion that we cannot derive a new expression. Therefore, Vakhidov's recommendation [6] of calculating Hansen coefficients of small eccentricities with Tayler series of $e_0 = 0$ has no practical importance.

Therefore, using the Tayler series expansion method, we must adopt the formulae of $e^* \neq 0$. In this way, the calculation formulae are complex. Moreover, as it is necessary to calculate the derivatives of Hansen coefficients, the formulae are more complex and the computing time is long. In addition, there are other methods (such as direct calculation methods based on Hill expressions of Hansen coefficients) of calculating higher-order Hansen coefficients in the cases of large eccentricity, and we thus do not recommend this method in view of its computational efficiency, although Wnuk [8] regarded it to be an advance in analytical orbit theory.

5.3 Interpolation Calculation and Chebyshev Polynomial Approximation

To reduce the computing time, Vakhidov studied polynomial approximations of Hansen coefficients on an interval of the eccentricity. These approximations included the Taylor series expansion stated in Sect. 5.2, Lagrange interpolation and Chebyshev polynomial approximation [6].

As the studied interval is too large, Vakhidov adopted a long word length calculation (where there are 32 digits) for construction of the interpolating polynomial and concluded that Lagrange interpolation on middle intervals of eccentricity has a high accuracy. Moreover, the required degree of the interpolating polynomial in this case should be set in the range of 15–20 [6].

For Chebyshev polynomial approximation, Vakhidov recommended the adoption of an approximation on a large interval of the eccentricity, rather than the adoption of approximations on many short intervals. Using Chebyshev polynomials with degree between 40 and 50, Vakhidov constructed Chebyshev polynomial approximations of Hansen coefficients on the interval $0 < e < 0.9$ efficiently [6].

N-point interpolation and N-degree polynomial approximation require the calculation of Hansen coefficients (and their derivatives) N times with the direct calculation method. The purpose of expressing Hansen coefficients on an interval of the eccentricity e is to deal with the situation that Hansen coefficients are required to be iteratively calculated many times in one orbit determination, but there are generally fewer than 10 iterations. Therefore, neither the use of the interpolation method nor the adoption of the Chebyshev polynomial approximation with N greater than 10 can save the computing time. It is then best that $N = 2$, and the interpolation method with $N = 2$ is introduced below.

If the values of the function $f(x)$ and its derivative at two nodes x_0 and x_1, i.e., $f(x_0)$, $f(x_1)$, $f'(x_0)$ and $f'(x_1)$, are known, the interpolating polynomial can be chosen as

$$\begin{aligned} Q(x) = {} & \frac{x_1 - x}{x_1 - x_0}\left[f(x_0) + \frac{1}{2}(x - x_0)f'(x_0)\right] \\ & + \frac{x - x_0}{x_1 - x_0}\left[f(x_1) + \frac{1}{2}(x - x_1)f'(x_1)\right] \end{aligned} \tag{5.25}$$

Evidently, when $f(e) = X_k^{n,m}(e)$, Hansen coefficients can also be calculated using the above formula. We carried out tests in which two groups of Hansen coefficients and their derivatives corresponding to two eccentricities e_0 and e_1 (with the difference between these two eccentricities being 0.02) were first calculated and, for the given $e_i = e_0 + 0.002i$ $(i = 1, 2, \cdots, 9)$ on the interval $[e_0, e_1]$, the corresponding Hansen coefficients of e_i were then calculated using the direct calculation method and compared with the results obtained by the interpolation of Eq. (5.25). The results are compared in Tables 5.5 and 5.6, in which $f(e) = K_{l-2p+q}^{-(l+1),l-2p}(e) = e^{-|q|}X_{l-2p+q}^{-(l+1),l-2p}(e)$.

Tables 5.5 and 5.6 respectively reveal that the interpolation results have five and six significant digits on the whole. The results for the large eccentricities are slightly worse, but they meet the requirements of first-order theory. Of course, for the interpolation calculation of the derivatives of Hansen coefficients, the second derivatives of Hansen coefficients at two eccentricities e_0 and e_1 are further required.

It is noted that the premise of this algorithm is that in one orbit determination, the correction of the eccentricity is less than 0.01. For precise orbit determination of the satellites, this requirement is not high and is achievable. If the requirement is in fact not possible, we can broaden the interval of the eccentricity $[e_0, e_1]$.

It is noted that the results are much worse when Hansen coefficients $X_{l-2p+q}^{-(l+1),l-2p}(e)$ are directly interpolated using the algorithm. The corresponding

Table 5.5 Interpolation test of $K_{l-2p+q}^{-(l+1),l-2p}(e)$ $(e \approx 0.7)$

l	p	q	e	Calculation with Eq. (3.67)	Interpolation calculation	Relative error
2	1	−6	0.702	9.95280445732E+00	9.95272948735E+00	7.53254747468E−06
2	1	−5	0.702	8.10111737459E+00	8.10104489484E+00	8.94688292753E−06
2	1	−4	0.702	6.57458370173E+00	6.57451350278E+00	1.06773222937E−05
2	1	−3	0.702	5.31672088217E+00	5.31665285848E+00	1.27942945451E−05
2	1	−2	0.702	4.28104085409E+00	4.28097498460E+00	1.53863252668E−05
2	1	−1	0.702	3.43127441704E+00	3.43121074109E+00	1.85575214370E−05
2	1	0	0.702	2.76844736082E+00	2.76838595212E+00	2.21816359530E−05
2	1	−6	0.704	9.97301715741E+00	9.97291500416E+00	1.02429636838E−05
2	1	−5	0.704	8.12541613408E+00	8.12531734823E+00	1.21576354764E−05
2	1	−4	0.704	6.60084422466E+00	6.60074852948E+00	1.44974147431E−05
2	1	−3	0.704	5.34343067734E+00	5.34333793382E+00	1.73565501572E−05
2	1	−2	0.704	4.30715650297E+00	4.30706668449E+00	2.08533131192E−05
2	1	−1	0.704	3.45611500369E+00	3.45602816418E+00	2.51263388551E−05
2	1	0	0.704	2.79163122761E+00	2.79154746697E+00	3.00041951193E−05
2	1	−6	0.706	9.99372519000E+00	9.99363255949E+00	9.26886760143E−06
2	1	−5	0.706	8.15019280548E+00	8.15010319110E+00	1.09953695153E−05
2	1	−4	0.706	6.62756749404E+00	6.62748065733E+00	1.31023504223E−05
2	1	−3	0.706	5.37058796834E+00	5.37050379017E+00	1.56739195063E−05
2	1	−2	0.706	4.33370309138E+00	4.33362155062E+00	1.88154925484E−05
2	1	−1	0.706	3.48136822909E+00	3.48128937511E+00	2.26502830682E−05
2	1	0	0.706	2.81520784551E+00	2.81513176811E+00	2.70237244916E−05
2	1	−6	0.708	1.00149399591E+01	1.00148821533E+01	5.77195076055E−06
2	1	−5	0.708	8.17545840117E+00	8.17540242345E+00	6.84704391619E−06
2	1	−4	0.708	6.65476416678E+00	6.65470988632E+00	8.15663242827E−06
2	1	−3	0.708	5.39820307503E+00	5.39815042753E+00	9.75278305254E−06
2	1	−2	0.708	4.36069060639E+00	4.36063958298E+00	1.17007631764E−05
2	1	−1	0.708	3.50704374172E+00	3.50699437391E+00	1.40767599000E−05
2	1	0	0.708	2.83918651290E+00	2.83913885556E+00	1.67855626106E−05
2	1	−6	0.710	1.00366732038E+01	1.00366637857E+01	9.38374923952E−07
2	1	−5	0.710	8.20122426152E+00	8.20121504527E+00	1.12376521017E−06
2	1	−4	0.710	6.68244522016E+00	6.68243621645E+00	1.34736768531E−06
2	1	−3	0.710	5.42628662978E+00	5.42627784589E+00	1.61876514615E−06
2	1	−2	0.710	4.38812933943E+00	4.38812078158E+00	1.95022656540E−06
2	1	−1	0.710	3.53315148603E+00	3.53314316056E+00	2.35638505102E−06
2	1	0	0.710	2.86357681550E+00	2.86356872930E+00	2.82381108087E−06
2	1	−6	0.712	1.00589370113E+01	1.00589774565E+01	4.02082681933E−06
2	1	−5	0.712	8.22750206717E+00	8.22754105657E+00	4.73891175237E−06

(continued)

Table 5.5 (continued)

l	p	q	e	Calculation with Eq. (3.67)	Interpolation calculation	Relative error
2	1	−4	0.712	6.71062196381E+00	6.71065964773E+00	5.61556268848E−06
2	1	−3	0.712	5.45484958910E+00	5.45488604526E+00	6.68325640273E−06
2	1	−2	0.712	4.41602989770E+00	4.41606514642E+00	7.98199121707E−06
2	1	−1	0.712	3.55970171362E+00	3.55973573507E+00	9.55739028064E−06
2	1	0	0.712	2.88838863733E+00	2.88842138935E+00	1.13392013935E−05
2	1	−6	0.714	1.00817438291E+01	1.00818231659E+01	7.86934564771E−06
2	1	−5	0.714	8.25430385177E+00	8.25438045735E+00	9.28068331414E−06
2	1	−4	0.714	6.73930605226E+00	6.73938018016E+00	1.09993366829E−05
2	1	−3	0.714	5.48390324602E+00	5.48397502564E+00	1.30891475726E−05
2	1	−2	0.714	4.44440321624E+00	4.44447267749E+00	1.56289260504E−05
2	1	−1	0.714	3.58670499499E+00	3.58677209744E+00	1.87086632486E−05
2	1	0	0.714	2.91363217212E+00	2.91369683570E+00	2.21934626074E−05
2	1	−6	0.716	1.01051064796E+01	1.01052009137E+01	9.34518941184E−06
2	1	−5	0.716	8.28164201540E+00	8.28173324760E+00	1.10161969285E−05
2	1	−4	0.716	6.76850949802E+00	6.76859781372E+00	1.30480286782E−05
2	1	−3	0.716	5.51345924282E+00	5.51354478702E+00	1.55155225093E−05
2	1	−2	0.716	4.47326057039E+00	4.47334337479E+00	1.85109715387E−05
2	1	−1	0.716	3.61417223180E+00	3.61425224767E+00	2.21394729357E−05
2	1	0	0.716	2.93931793528E+00	2.93939506836E+00	2.62418282714E−05
2	1	−6	0.718	1.01290381735E+01	1.01291107001E+01	7.16027202911E−06
2	1	−5	0.718	8.30952933852E+00	8.30959942733E+00	8.43475173471E−06
2	1	−4	0.718	6.79824468529E+00	6.79831254844E+00	9.98245099045E−06
2	1	−3	0.718	5.54352958450E+00	5.54359532940E+00	1.18597553761E−05
2	1	−2	0.718	4.50261358901E+00	4.50267723834E+00	1.41360844304E−05
2	1	−1	0.718	3.64211466969E+00	3.64217618575E+00	1.68902039049E−05
2	1	0	0.718	2.96545677645E+00	2.96551608731E+00	2.00005814197E−05

calculation results for the same test reported in Table 5.6 are listed in Table 5.7. The interpolation accuracy of $X_{l-2p+q}^{-(l+1),l-2p}(e)$ is almost four orders of magnitude worse. The reason for the worse performance is that the first term (also the main term) of $K_{l-2p+q}^{-(l+1),l-2p}(e)$ does not contain e, and the change on the interval $[e_0, e_1]$ is relatively gentle and can be approximated by low-order polynomials. However, $X_{l-2p+q}^{-(l+1),l-2p}(e)$ cannot be approximated using low-order polynomials. Therefore, the interpolation accuracies of $K_{l-2p+q}^{-(l+1),l-2p}(e)$ and $X_{l-2p+q}^{-(l+1),l-2p}(e)$ differ greatly.

In this way, the calculation of $K_k^{n,m}(e) = e^{-|k-m|}X_k^{n,m}(e)$ is not only a requirement of theory without singularities but also improves the accuracy of the interpolation method in calculating Hansen coefficients $X_k^{n,m}(e)$. However, the direct calculation

Table 5.6 Interpolation test of $K_{l-2p+q}^{-(l+1),l-2p}(e)$ $(e \approx 0.1)$

l	p	q	e	Calculation with Eq. (3.67)	Interpolation calculation	Relative error
2	0	6	0.092	9.93225255143E+01	9.93224835237E+01	4.22770459248E−07
2	0	7	0.092	1.64820041720E+02	1.64819958870E+02	5.02669271743E−07
2	0	8	0.092	2.67209768919E+02	2.67209611453E+02	5.89297177043E−07
2	0	9	0.092	4.25269927654E+02	4.25269637362E+02	6.82605102005E−07
2	0	10	0.092	6.66730204080E+02	6.66729682332E+02	7.82547480142E−07
2	0	6	0.094	9.92103951398E+01	9.92103383554E+01	5.72363572321E−07
2	0	7	0.094	1.64620118412E+02	1.64620006388E+02	6.80501155647E−07
2	0	8	0.094	2.66862706216E+02	2.66862493329E+02	7.97739805839E−07
2	0	9	0.094	4.24680502099E+02	4.24680109689E+02	9.24012188128E−07
2	0	10	0.094	6.65747358399E+02	6.65746653202E+02	1.05925572345E−06
2	0	6	0.096	9.90959581382E+01	9.90959072847E+01	5.13174263549E−07
2	0	7	0.096	1.64416102881E+02	1.64416002579E+02	6.10049411055E−07
2	0	8	0.096	2.66508574498E+02	2.66508383929E+02	7.15059383811E−07
2	0	9	0.096	4.24079129986E+02	4.24078778788E+02	8.28141101869E−07
2	0	10	0.096	6.64744689880E+02	6.64744058881E+02	9.49235762189E−07
2	0	6	0.098	9.89792211121E+01	9.89791903115E+01	3.11182849226E−07
2	0	7	0.098	1.64208008161E+02	1.64207947443E+02	3.69764486584E−07
2	0	8	0.098	2.66147398552E+02	2.66147283251E+02	4.33225369455E−07
2	0	9	0.098	4.23465857037E+02	4.23465644660E+02	5.01521290626E−07
2	0	10	0.098	6.63722280752E+02	6.63721899370E+02	5.74610647320E−07
2	0	6	0.100	9.88601907945E+01	9.88601874358E+01	3.39742717094E−08
2	0	7	0.100	1.63995847543E+02	1.63995840980E+02	4.00177224523E−08
2	0	8	0.100	2.65779203648E+02	2.65779191295E+02	4.64751985302E−08
2	0	9	0.100	4.22840729852E+02	4.22840707303E+02	5.33285292691E−08
2	0	10	0.100	6.62680214799E+02	6.62680174668E+02	6.05598622376E−08
2	0	6	0.102	9.87388740478E+01	9.87388986577E+01	2.49242167945E−07
2	0	7	0.102	1.63779634569E+02	1.63779683191E+02	2.96873801955E−07
2	0	8	0.102	2.65404015529E+02	2.65404108063E+02	3.48654910084E−07
2	0	9	0.102	4.22203795904E+02	4.22203966718E+02	4.04577372023E−07
2	0	10	0.102	6.61618577364E+02	6.61618884775E+02	4.64635085714E−07
2	0	6	0.104	9.86152778633E+01	9.86153239771E+01	4.67612576047E−07
2	0	7	0.104	1.63559383034E+02	1.63559474075E+02	5.56621801378E−07
2	0	8	0.104	2.65021860415E+02	2.65022033553E+02	6.53298358958E−07
2	0	9	0.104	4.21555103529E+02	4.21555422905E+02	7.57613892779E−07
2	0	10	0.104	6.60537455325E+02	6.60538029691E+02	8.69543874604E−07
2	0	6	0.106	9.84894093610E+01	9.84894633940E+01	5.48617694363E−07
2	0	7	0.106	1.63335106983E+02	1.63335213631E+02	6.52940638877E−07

(continued)

Table 5.6 (continued)

l	p	q	e	Calculation with Eq. (3.67)	Interpolation calculation	Relative error
2	0	8	0.106	2.64632764998E+02	2.64632967766E+02	7.66225724455E−07
2	0	9	0.106	4.20894701927E+02	4.20895075865E+02	8.88436072917E−07
2	0	10	0.106	6.59436937095E+02	6.59437609417E+02	1.01953931019E−06
2	0	6	0.108	9.83612757884E+01	9.83613169085E+01	4.18051623435E−07
2	0	7	0.108	1.63106820713E+02	1.63106901862E+02	4.97520515675E−07
2	0	8	0.108	2.64236756438E+02	2.64236910702E+02	5.83811502605E−07
2	0	9	0.108	4.20222641149E+02	4.20222925596E+02	6.76895795759E−07
2	0	10	0.108	6.58317112606E+02	6.58317623952E+02	7.76748045470E−07

methods based on Hill expressions of Hansen coefficients cannot be used to calculate $K_k^{n,m}(e)$, and other direct calculation methods are not as stable as the direct calculation methods based on Hill expressions. Therefore, further research on the calculation of $K_k^{n,m}(e)$ using direct methods based on Hill expressions is needed.

Besides the above methods, Klioner et al. studied the use of the fast Fourier transform (FFT) in calculating Hansen coefficients [9]. In addition, Lion and Métris investigated Hansen-like coefficients based on the eccentric anomaly and their calculation methods [10], to avert difficulties in calculating Hansen coefficients. However, these methods are not mature and have not yet been applied. Hence, their details are not presented here.

Table 5.7 Interpolation test of Hansen coefficients $X_{l-2p+q}^{-(l+1),l-2p}(e)$ $(e \approx 0.1)$

l	p	q	e	Calculation with Eq. (3.17) or Eq. (3.87)	Interpolation calculation	Relative error
2	0	−10	0.106	5.16189326796E−11	5.27299491662E−11	2.15234300466E−02
2	0	−9	0.106	3.66861847726E−10	3.72870396205E−10	1.63782320692E−02
2	0	−8	0.106	2.59855384549E−09	2.62942207239E−09	1.18790022229E−02
2	0	−7	0.106	1.82805506414E−08	1.84283771325E−08	8.08654476485E−03
2	0	−6	0.106	1.26806854093E−07	1.27446178864E−07	5.04172093036E−03
2	0	−5	0.106	8.52850406556E−07	8.55202953396E−07	2.75845191914E−03
2	0	−4	0.106	5.30205984487E−06	5.30850051259E−06	1.21474821333E−03
2	0	−3	0.106	2.50060512129E−05	2.50145851785E−05	3.41276019940E−04
2	0	1	0.106	3.61895059599E−01	3.61892245635E−01	7.77563706720E−06
2	0	2	0.106	9.31039506592E−02	9.31011732688E−02	2.98310689922E−05

(continued)

Table 5.7 (continued)

l	p	q	e	Calculation with Eq. (3.17) or Eq. (3.87)	Interpolation calculation	Relative error
2	0	3	0.106	2.04051747520E−02	2.04104336820E−02	2.57725308177E−04
2	0	4	0.106	4.08432903692E−03	4.08871138501E−03	1.07296646457E−03
2	0	5	0.106	7.71029339885E−04	7.72997286456E−04	2.55236275661E−03
2	0	6	0.106	1.39709109533E−04	1.40375434932E−04	4.76937689557E−03
2	0	7	0.106	2.45595609216E−05	2.47498771287E−05	7.74916977172E−03
2	0	8	0.106	4.21784422950E−06	4.26626600158E−06	1.14802181986E−02
2	0	9	0.106	7.11092742861E−07	7.22415791306E−07	1.59234481846E−02
2	0	10	0.106	1.18095111981E−07	1.20577396514E−07	2.10193672828E−02
2	0	−10	0.108	6.22224965281E−11	6.31711703414E−11	1.52464762147E−02
2	0	−9	0.108	4.34070886551E−10	4.39092298525E−10	1.15681842052E−02
2	0	−8	0.108	3.01791869399E−09	3.04319554294E−09	8.37558977179E−03
2	0	−7	0.108	2.08391527594E−08	2.09578902939E−08	5.69781007323E−03
2	0	−6	0.108	1.41887645317E−07	1.42391878316E−07	3.55374844516E−03
2	0	−5	0.108	9.36655609416E−07	9.38479341578E−07	1.94706799743E−03
2	0	−4	0.108	5.71541266884E−06	5.72032539468E−06	8.59557503341E−04
2	0	−3	0.108	2.64562121171E−05	2.64626260973E−05	2.42437585500E−04
2	0	1	0.108	3.68371949556E−01	3.68369842451E−01	5.72004802060E−06
2	0	2	0.108	9.65561921373E−02	9.65540784044E−02	2.18912208961E−05
2	0	3	0.108	2.15596964303E−02	2.15636048198E−02	1.81282213269E−04
2	0	4	0.108	4.39649010622E−03	4.39981359778E−03	7.55942008526E−04
2	0	5	0.108	8.45542901815E−04	8.47061882053E−04	1.79645554987E−03
2	0	6	0.108	1.56086982921E−04	1.56610553659E−04	3.35435235177E−03
2	0	7	0.108	2.79536427741E−05	2.81059856151E−05	5.44983858542E−03
2	0	8	0.108	4.89083795158E−06	4.93035911997E−06	8.08065382304E−03
2	0	9	0.108	8.40027004071E−07	8.49459535441E−07	1.12288430305E−02
2	0	10	0.108	1.42125727054E−07	1.44238553048E−07	1.48658940118E−02

References

1. Sharaf MA, Selim HH (2010) Recursive harmonic analysis for computing Hansen coefficients. Res Astron Astrophys 10(12):1928–1306
2. Wu LD (2011) Orbit and detection of artificial satellites and space debris. China Science and Technology Press, Beijing (In Chinese)
3. de Moraes RV, Wnuk E (1998) Orbital perturbations using geopotential coefficients up to high degree and order: highly eccentric orbits. In: Spaceflight dynamics, vol 100, Part 1 and 2. Univelt Inc., San Diego, v 100, pp 227–238
4. Giacaglia GEO (1976) A note on Hansen's coefficients in satellite theory. Celest Mech 14(4):515–523

5. Jarnagin MP (1965) Expansions in elliptic motion. In: Astronomical papers prepared for the use of the American Ephemeris and Nautical Almanac Volume XVIII, U.S. Government Printing Office, Washington
6. Vakhidov AA (2000) Construction of polynomial approximation for Hansen coefficients. Comput Phys Commun 124(1):40–48
7. Da Silva Fernandes S (1995) Some expansions of the elliptic motion to high eccentricities. Celest Mech Dyn Astron 62(4):305–321
8. Wnuk E (1999) Recent progress in analytical orbit theories. Adv Space Res 23(4):677–687
9. Klioner SA, Vakhidov AA, Vasiliev NN (1998) Numerical computation of Hansen-like expansions. Celest Mech Dyn Astron 68(3):257–272
10. Lion G, Métris G (2013) Two algorithms to compute Hansen-like coefficients with respect to the eccentric anomaly. Adv Space Res 51(1):1–9

Appendix A
Deduction of Three Formulae

A.1 Deduction of Eq. (2.5)

Laskar and Boué presented the series expression of Hansen coefficients $X_0^{n,m}$ [1]:

$$X_0^{n,m} = (-1)^m \frac{(n+1+m)!}{(n+1)!} \sum_{l=0}^{[(n+1-m)/2]} \frac{(n+1-m)!}{l!(m+l)!(n+1-m-2l)!} \left(\frac{e}{2}\right)^{m+2l} \tag{A1}$$

Equation (A1) can be proved through mathematical induction.

First, it is proved that Eq. (A1) holds when $m = 0$ and $m = 1$. According to the definition of $X_0^{n,m}$, we have

$$X_0^{n,m} = \frac{1}{2\pi} \int_0^{2\pi} \left(\frac{r}{a}\right)^n \exp(jmf)\mathrm{d}M = \frac{1}{2\pi} \int_0^{2\pi} \left(\frac{r}{a}\right)^{n+1} \exp(jmf)\mathrm{d}E \tag{A2}$$

when $m = 0$, we have

$$X_0^{n,0} = \frac{1}{2\pi} \int_0^{2\pi} (1 - e\cos E)^{n+1}\mathrm{d}E \tag{A3}$$

and when $m = 1$, we have

$$X_0^{n,1} = \frac{1}{2\pi} \int_0^{2\pi} (1 - e\cos E)^n(\cos E - e)\mathrm{d}E \tag{A4}$$

L.D Wu and M.J Zhang, *Hansen Coefficients in Satellite Orbital Dynamics*,
Springer Aerospace Technology, https://doi.org/10.1007/978-981-97-0456-9

Equation (A3) can be further expressed as

$$X_0^{n,0} = \frac{1}{2\pi} \int_0^{2\pi} \sum_{k=0}^{n+1} \binom{n+1}{k} (-e \cos E)^k \mathrm{d}E \tag{A5}$$

Substituting $\cos^k E = \frac{1}{2^k} \sum_{l=0}^{k} \binom{k}{l} \cos(k-2l)E$ into Eq. (A5), we obtain

$$X_0^{n,0} = \frac{1}{2\pi} \int_0^{2\pi} \sum_{k=0}^{n+1} \binom{n+1}{k} \left(-\frac{e}{2}\right)^k \sum_{l=0}^{k} \binom{k}{l} \cos(k-2l) E \mathrm{d}E \tag{A6}$$

Only if $2l = k$, i.e., k is even, the terms in Eq. (A6) are not equal to zero. Letting $k = 2l$, we immediately obtained

$$X_0^{n,0} = \sum_{l=0}^{[(n+1)/2]} \binom{n+1}{2l} \left(-\frac{e}{2}\right)^{2l} \binom{2l}{l} = \sum_{l=0}^{[(n+1)/2]} \frac{(n+1)!}{l!l!(n+1-2l)!} \left(\frac{e}{2}\right)^{2l} \tag{A7}$$

Equation (A1) then holds when $m = 0$.

Similarly,

$$\begin{aligned} X_0^{n,1} &= \frac{1}{2\pi} \int_0^{2\pi} (1 - e\cos E)^n (\cos E - e) \mathrm{d}E \\ &= \frac{1}{2\pi} \int_0^{2\pi} \sum_{k=0}^{n} \binom{n}{k} \left(-\frac{e}{2}\right)^k \sum_{l=0}^{k} \binom{k}{l} \cos(k-2l) E (\cos E - e) \mathrm{d}E \end{aligned} \tag{A8}$$

In Eq. (A8), one part is

$$\begin{aligned} &-\frac{1}{2\pi} \int_0^{2\pi} (1 - e\cos E)^n e \mathrm{d}E \\ &\quad = -\frac{e}{2\pi} \int_0^{2\pi} \sum_{k=0}^{n} \binom{n}{k} \left(-\frac{e}{2}\right)^k \sum_{l=0}^{k} \binom{k}{l} \cos(k-2l) E \mathrm{d}E \\ &\quad = -2 \sum_{l=0}^{[n/2]} \frac{n!}{l!l!(n-2l)!} \left(\frac{e}{2}\right)^{2l+1} \end{aligned} \tag{A9}$$

and the other part is

$$
\begin{aligned}
&\frac{1}{2\pi}\int_0^{2\pi}(1-e\cos E)^n\cos E\mathrm{d}E \\
&\quad=\frac{1}{2\pi}\int_0^{2\pi}\sum_{k=0}^{n}\binom{n}{k}\left(-\frac{e}{2}\right)^k\sum_{l=0}^{k}\binom{k}{l}\cos(k-2l)E\cos E\mathrm{d}E
\end{aligned}
\tag{A10}
$$

Only if $k = 2l \pm 1$, the terms in Eq. (A10) are not equal to zero. As the terms with $k = 2l + 1$ and $k = 2l - 1$ are equal, we let $k = 2l + 1$ in Eq. (A10), and we multiply the result by 2. Equation (A10) then becomes

$$
\begin{aligned}
&\frac{1}{2\pi}\int_0^{2\pi}\sum_{k=0}^{n}\binom{n}{k}\left(-\frac{e}{2}\right)^k\sum_{l=0}^{k}\binom{k}{l}\cos(k-2l)E\cos E\mathrm{d}E \\
&\qquad=-\sum_{l=0}^{[n/2]}\frac{n!}{l!(l+1)!(n-2l-1)!}\left(\frac{e}{2}\right)^{2l+1}
\end{aligned}
\tag{A11}
$$

Adding Eqs. (A9) and (A11), we obtain

$$
\begin{aligned}
X_0^{n,1}&=-\sum_{l=0}^{[n/2]}[(2l+2)+(n-2l)]\frac{n!}{l!(l+1)!(n-2l)!}\left(\frac{e}{2}\right)^{2l+1} \\
&=-(n+2)\sum_{l=0}^{[[n/2]]}\frac{n!}{l!(l+1)!(n-2l)!}\left(\frac{e}{2}\right)^{2l+1}
\end{aligned}
\tag{A12}
$$

Equation (A1) then holds when $m = 1$.

Next, based on the recursion formula

$$
X_0^{n,m}=\frac{1}{n+2-m}\left[\frac{2}{e}(m-1)X_0^{n,m-1}+(n+m)X_0^{n,m-2}\right]
\tag{A13}
$$

we prove Eq. (A1) through mathematical induction.

Proof:

(1) Equation (A1) holds when $m = 0$ and $m = 1$.
(2) If Eq. (A1) holds, letting $m = m - 1$, we have

$$
X_0^{n,m-1}=(-1)^{m-1}\frac{(n+1+m-1)!}{(n+1)!}
$$

$$\sum_{l=0}^{[(n+1-m+1)/2]} \frac{(n+1-m+1)!}{l!(m-1+l)!(n+1-m+1-2l)!}\left(\frac{e}{2}\right)^{m-1+2l}$$

Letting $m = m - 2$, we have

$$X_0^{n,m-2} = (-1)^{m-2}\frac{(n+1+m-2)!}{(n+1)!}$$
$$\sum_{l=0}^{[(n+1-m+2)/2]} \frac{(n+1-m+2)!}{l!(m-2+l)!(n+1-m+2-2l)!}\left(\frac{e}{2}\right)^{m-2+2l}$$

According to the recursion formula Eq. (A13), we have

$$\begin{aligned}
X_0^{n,m} &= \frac{1}{n+2-m}\left[\frac{2}{e}(m-1)X_0^{n,m-1} + (n+m)X_0^{n,m-2}\right] \\
&= \frac{(-1)^{m-1}}{n+2-m}\frac{(n+m)!}{(n+1)!} \\
&\quad \sum_{l=0}^{[(n-m+2)/2]} \frac{(m-1)(n-m+2)!}{l!(m-1+l)!(n-m+2-2l)!}\left(\frac{e}{2}\right)^{m-2+2l} \\
&\quad + \frac{(-1)^{m-2}}{n+2-m}\frac{(n+m-1)!}{(n+1)!} \\
&\quad \sum_{l=0}^{[(n-m+3)/2]} \frac{(n+m)(n-m+3)!}{l!(m-2+l)!(n-m+3-2l)!}\left(\frac{e}{2}\right)^{m-2+2l} \\
&= (-1)^{m-1}\frac{(n+m+1)!}{(n+1)!} \\
&\quad \sum_{l=0}^{[(n-m+2)/2]} \frac{(m-1)(n-m+1)!}{(n+m+1)l!(m-1+l)!(n-m+2-2l)!}\left(\frac{e}{2}\right)^{m-2+2l} \\
&\quad + (-1)^{m-2}\frac{(n+m+1)!}{(n+1)!} \\
&\quad \sum_{l=0}^{[(n-m+3)/2]} \frac{(n-m+3)(n-m+1)!}{(n+m+1)l!(m-2+l)!(n-m+3-2l)!}\left(\frac{e}{2}\right)^{m-2+2l}
\end{aligned}$$

It is clear that the term with $l = 0$ in the above formula can be eliminated. Replacing l with $l + 1$, i.e., $l \Rightarrow l + 1$, the above formula becomes

$$X_0^{n,m} = (-1)^{m-1}\frac{(n+m+1)!}{(n+1)!}$$

$$
\begin{aligned}
&\times \sum_{l=0}^{[(n-m)/2]} \frac{(m-1)}{(n+m+1)(l+1)} \frac{(n-m+1)!}{l!(m+l)!(n-m-2l)!}\left(\frac{e}{2}\right)^{m+2l} \\
&+(-1)^{m-2}\frac{(n+m+1)!}{(n+1)!} \\
&\times \sum_{l=0}^{[(n-m+1)/2]} \frac{(n-m+3)}{(n+m+1)(l+1)} \frac{(n-m+1)!}{l!(m+l-1)!(n-m+1-2l)!}\left(\frac{e}{2}\right)^{m+2l} \\
=\;&(-1)^{m-1}\frac{(n+m+1)!}{(n+1)!} \\
&\times \sum_{l=0}^{[(n-m)/2]} \frac{(m-1)(n-m+1-2l)}{(n+m+1)(l+1)} \frac{(n-m+1)!}{l!(m+l)!(n-m+1-2l)!}\left(\frac{e}{2}\right)^{m+2l} \\
&+(-1)^{m-2}\frac{(n+m+1)!}{(n+1)!} \\
&\times \sum_{l=0}^{[(n-m+1)/2]} \frac{(n-m+3)(m+l)}{(n+m+1)(l+1)} \frac{(n-m+1)!}{l!(m+l)!(n-m+1-2l)!}\left(\frac{e}{2}\right)^{m+2l} \\
=\;&(-1)^{m}\frac{(n+m+1)!}{(n+1)!} \\
&\times \sum_{l=0}^{[(n-m)/2]} \left[\frac{(n-m+3)(m+l)}{(n+m+1)(l+1)} - \frac{(m-1)(n-m+1-2l)}{(n+m+1)(l+1)}\right] \\
&\times \frac{(n-m+1)!}{l!(m+l)!(n-m+1-2l)!}\left(\frac{e}{2}\right)^{m+2l} \\
&+(-1)^{m}\frac{(n+m+1)!}{(n+1)!} \\
&\times \frac{(n-m+3)(m+l)}{(n+m+1)(l+1)} \frac{(n-m+1)!}{l!(m+l)!(n-m+1-2l)!}\left(\frac{e}{2}\right)^{m+2l}\Bigg|_{l=[(n-m+1)/2]}
\end{aligned} \tag{A14}
$$

When $n-m$ is even, the last term in Eq. (A14) can be eliminated. As

$$
\frac{(n-m+3)(m+l)}{(n+m+1)(l+1)} - \frac{(m-1)(n-m+1-2l)}{(n+m+1)(l+1)} = 1
$$

from Eq. (A14), we obtain

$$
X_0^{n,m} = (-1)^m \frac{(n+m+1)!}{(n+1)!} \sum_{l=0}^{[(n-m)/2]} \frac{(n-m+1)!}{l!(m+l)!(n-m+1-2l)!}\left(\frac{e}{2}\right)^{m+2l} \tag{A15}
$$

When $n-m$ is odd, the last term in Eq. (A14) should be added. As

$$\frac{(n-m+3)(m+l)}{(n+m+1)(l+1)}\Big|_{l=(n-m+1)/2}=\frac{(n-m+3)(n+m+1)/2}{(n+m+1)(n-m+3)/2}=1$$

from Eq. (A14), we obtain

$$X_0^{n,m}=(-1)^m\frac{(n+m+1)!}{(n+1)!}\sum_{l=0}^{[(n+1-m)/2]}\frac{(n-m+1)!}{l!(m+l)!(n-m+1-2l)!}\left(\frac{e}{2}\right)^{m+2l} \tag{A16}$$

Equation (A1), i.e., Eq. (2.5), is thus proven.

A.2 Deduction of Eq. (3.74)

Challe and Laclaverie introduced a type of Hansen coefficients defined as [2]

$$\begin{aligned}B_{n,p}^k(e)&=\frac{1}{2\pi}\int_0^{2\pi}\left(\frac{a}{r}\right)^n\exp[j(kf-pM)]\mathrm{d}M\\&=\frac{1}{2\pi}\int_0^{2\pi}\left(\frac{a}{r}\right)^{n-1}\exp[j(kf-pM)]\mathrm{d}E\end{aligned} \tag{A17}$$

and presented the recursion formulae [2]:

$$B_{n,p}^k=\frac{1}{\eta^2}\left[B_{n-1,p}^k+\frac{e}{2}\left(B_{n-1,p}^{k+1}+B_{n-1,p}^{k-1}\right)\right] \tag{A18}$$

$$p\eta B_{n,p}^k=k\eta^2B_{n+2,p}^k+\frac{ne}{2}\left(B_{n+1,p}^{k+1}-B_{n+1,p}^{k-1}\right) \tag{A19}$$

$$B_{n,p}^k=\frac{1}{p\eta}\left[kB_{n+1,p}^k+(n+k)\frac{e}{2}B_{n+1,p}^{k+1}-(n-k)\frac{e}{2}B_{n+1,p}^{k-1}\right] \tag{A20}$$

$$\begin{aligned}4\left[p\eta^3-k\left(1+\tfrac{e^2}{2}\right)\right]B_{n,p}^k&=e^2(k-n)B_{n,p}^{k-2}+2e(2k-n)B_{n,p}^{k-1}\\&+2e(2k+n)B_{n,p}^{k+1}+e^2(k+n)B_{n,p}^{k+2}\end{aligned} \tag{A21}$$

Under the premise that the orbital elements a, e, i, Ω, ω and E (where the mean anomaly M is replaced by the eccentric anomaly E) are independent variables, by differentiating Eq. (A17) with respect to e, we have

$$\frac{\mathrm{d}}{\mathrm{d}e}B_{n,p}^k=\frac{1}{2\pi}\int_0^{2\pi}(n-1)\left(\frac{a}{r}\right)^{n-2}\frac{\partial}{\partial e}\left(\frac{a}{r}\right)\exp[j(kf-pM)]\mathrm{d}E$$

$$+\frac{1}{2\pi}\int_0^{2\pi}\left(\frac{a}{r}\right)^{n-1}kj\frac{\partial f}{\partial e}\exp[j(kf-pM)]\mathrm{d}E$$
$$-\frac{1}{2\pi}\int_0^{2\pi}\left(\frac{a}{r}\right)^{n-1}pj\frac{\partial M}{\partial e}\exp[j(kf-pM)]\mathrm{d}E \tag{A22}$$

where

$$\begin{aligned}
\frac{\partial}{\partial e}\left(\frac{a}{r}\right) &= \left(\frac{a}{r}\right)^2\cos E = \frac{1}{e}\left(\frac{a^2}{r^2}-\frac{a}{r}\right)\\
\frac{\partial M}{\partial e} &= -\sin E = -\frac{r}{a}\frac{\sin f}{\eta} = -\frac{r}{a}\frac{1}{2j\eta}\left[\exp(jf)-\exp(-jf)\right]\\
\frac{\partial f}{\partial e} &= \frac{1}{\eta^2}\sin f = \frac{1}{2j\eta^2}\left[\exp(jf)-\exp(-jf)\right]
\end{aligned} \tag{A23}$$

The formula for the derivative $\frac{dB_{n,p}^k}{de}$ is then obtained as

$$\frac{\mathrm{d}B_{n,p}^k}{\mathrm{d}e} = \frac{n-1}{e}\left(B_{n+1,p}^k - B_{n,p}^k\right) + \frac{k}{2\eta^2}\left(B_{n,p}^{k+1} - B_{n,p}^{k-1}\right) + \frac{p}{2\eta}\left(B_{n-1,p}^{k+1} - B_{n-1,p}^{k-1}\right) \tag{A24}$$

Note that the original Eqs. (A21) and (A24) in Ref. [2] have errors.

Using Eqs. (A18) and (A20), Eq. (A24) becomes

$$\frac{\mathrm{d}B_{n,p}^k}{\mathrm{d}e} = \frac{1}{\eta^2}\left(\frac{n-k}{4}eB_{n,p}^{k-2} + \frac{n-2k}{2}B_{n,p}^{k-1} + \frac{ne}{2}B_{n,p}^k + \frac{n+2k}{2}B_{n,p}^{k+1} + \frac{n+k}{4}eB_{n,p}^{k+2}\right) \tag{A25}$$

From Eq. (A21), we have

$$\begin{aligned}
&(n-k)\frac{e^2}{4}B_{n,p}^{k-2} + (n-2k)\frac{e}{2}B_{n,p}^{k-1} + \left[p\eta^3 - k\left(1+\frac{e^2}{2}\right)\right]B_{n,p}^k\\
&\qquad = (n+k)\frac{e^2}{4}B_{n,p}^{k+2} + (n+2k)\frac{e}{2}B_{n,p}^{k+1}\\
&(n+k)\frac{e^2}{4}B_{n,p}^{k+2} + (n+2k)\frac{e}{2}B_{n,p}^{k+1} - \left[p\eta^3 - k\left(1+\frac{e^2}{2}\right)\right]B_{n,p}^k\\
&\qquad = (n-k)\frac{e^2}{4}B_{n,p}^{k-2} + (n-2k)\frac{e}{2}B_{n,p}^{k-1}
\end{aligned} \tag{A26}$$

Substituting Eq. (A26) into Eq. (A25), it is immediately obtained that

$$\frac{\mathrm{d}B_{n,p}^{k}}{\mathrm{d}e}=\frac{1}{\eta^2}\left\{\left[\frac{ne}{2}-\frac{p\eta^3}{e}+k\left(\frac{e}{2}+\frac{1}{e}\right)\right]B_{n,p}^{k}+(n+2k)B_{n,p}^{k+1}+\frac{n+k}{2}eB_{n,p}^{k+2}\right\} \tag{A27}$$

$$\frac{\mathrm{d}B_{n,p}^{k}}{\mathrm{d}e}=\frac{1}{\eta^2}\left\{\left[\frac{ne}{2}+\frac{p\eta^3}{e}-k\left(\frac{e}{2}+\frac{1}{e}\right)\right]B_{n,p}^{k}+(n-2k)B_{n,p}^{k-1}+\frac{n-k}{2}eB_{n,p}^{k-2}\right\} \tag{A28}$$

In addition, according to the definition of $B_{n,p}^{k}(e)$, we have

$$B_{n,p}^{k}(e)=X_{p}^{-n,k}(e) \tag{A29}$$

Letting $k\Rightarrow m$ and $p\Rightarrow k$, Eqs. (A27) and (A28) become

$$\begin{aligned}\frac{dX_{k}^{-n,m}}{de}&=\frac{1}{\eta^2}\left\{\left[\frac{ne}{2}-\frac{k\eta^3}{e}+m\left(\frac{e}{2}+\frac{1}{e}\right)\right]X_{k}^{-n,m}\right.\\&\left.\quad+(n+2m)X_{k}^{-n,m+1}+\frac{n+m}{2}eX_{k}^{-n,m+2}\right\}\end{aligned} \tag{A30}$$

$$\begin{aligned}\frac{dX_{k}^{-n,m}}{de}&=\frac{1}{\eta^2}\left\{\left[\frac{ne}{2}+\frac{k\eta^3}{e}-m\left(\frac{e}{2}+\frac{1}{e}\right)\right]X_{k}^{-n,m}\right.\\&\left.\quad+(n-2m)X_{k}^{-n,m-1}+\frac{n-m}{2}eX_{k}^{-n,m-2}\right\}\end{aligned} \tag{A31}$$

Replacing $-n$ in Eq. (A31) by n, we obtain

$$\begin{aligned}2\eta^2e\frac{\mathrm{d}X_{k}^{n,m}}{\mathrm{d}e}&=\left[-2m-(n+m)e^2+2k\eta^3\right]X_{k}^{n,m}\\&\quad-2(n+2m)eX_{k}^{n,m-1}-(n+m)e^2X_{k}^{n,m-2}\end{aligned} \tag{A32}$$

which is Eq. (3.74b). Similarly, replacing $-n$ in Eq. (A30) by n, Eq. (3.74a) is immediately obtained.

A.3 Train of Thought for the Deduction of Eq. (3.10)

Replacing the corresponding symbols with those in Ref. [3], Eq. (3.19) becomes

$$\begin{aligned}X_{k}^{n,m}&=\left(1+\beta^2\right)^{-n-1}\sum_{p}\sum_{q}\binom{n+m+1}{p}\binom{n-m+1}{q}\\&\quad(-\beta)^{p+q}J_{k-m+p-q}(ke)\end{aligned} \tag{A33}$$

As

$$\beta = \frac{e}{2}\left(1+\beta^2\right) \tag{A34}$$

$$(-\beta)^{p+q} = (-1)^{p+q}\left(\frac{e}{2}\right)^{p+q}\left(1+\beta^2\right)^{p+q} \tag{A35}$$

it follows from Eq. (A33) that

$$X_k^{n,m} = \left(1+\beta^2\right)^{-n+p+q-1}\sum_p\sum_q\binom{n+m+1}{p}\binom{n-m+1}{q}$$
$$(-1)^{p+q}\left(\frac{e}{2}\right)^{p+q}J_{k-m+p-q}(ke) \tag{A36}$$

Furthermore, using

$$\left(1+\beta^2\right)^{-n+p+q-1} = \sum_{j=0}^{\infty}\left[2\binom{-n+p+q+2j-2}{j}\right.$$
$$\left.-\binom{-n+p+q+2j-1}{j}\right]\left(\frac{e}{2}\right)^{2j}$$
$$J_n(x) = \sum_{i=0}^{\infty}\frac{(-1)^i}{i!(n+i)!}\left(\frac{x}{2}\right)^{n+2i} \tag{A37}$$

Eq. (A36) becomes

$$X_k^{n,m} = \sum_p\sum_q\sum_j\sum_\alpha\binom{n+m+1}{p}\binom{n-m+1}{q}$$
$$\times\left[2\binom{-n+p+q+2j-2}{j}-\binom{-n+p+q+2j-1}{j}\right]$$
$$\times(-1)^{p+q}\frac{(-1)^\alpha k^{k-m+p-q+2\alpha}}{\alpha!(k-m+p-q+\alpha)!}\left(\frac{e}{2}\right)^{k-m+2p+2\alpha+2j} \tag{A38}$$

which is Eq. (6-5) in Ref. [3].

The subsequent deduction is to transform and optimize the summation indexes, and eliminate some summation ranges where binomials are zero. Equation (3.10) is then obtained as

$$X_{m+s}^{n,m} = \left(-\frac{e}{2}\right)^s \sum_{t=0}^{\infty} \left(\frac{e}{2}\right)^{2t} \left\{ \sum_{j=0}^{t} \sum_{p=0}^{j} \binom{n+m+1}{j-p} X_p \sum_{q=0}^{s+j} \binom{n-m+1}{s+j-q} X_q \left[\binom{k}{t-j} - \binom{k}{t-j-1} \right] \right\}$$

where

$$k = 2t - n + s - p - q - 2, \quad X_p = \frac{(m+s)^p}{p!}, \quad X_q = \frac{(m+s)^q}{q!}(-1)^q$$

This deduction process is complex. Interested readers can try further deduction.

Appendix B
Hansen Expressions of Hansen Coefficients

Let

$$x = \exp(jf), \quad y = \exp(jE), \quad z = \exp(jM) \tag{B1}$$

where f, E and M are the true anomaly, eccentric anomaly and mean anomaly respectively. Substituting the relation

$$\mathrm{d}M = \frac{\mathrm{d}z}{jz} \tag{B2}$$

into the definition of Hansen coefficients

$$X_k^{n,s} = \frac{1}{2\pi}\int_0^{2\pi}\left(\frac{r}{a}\right)^n \exp[j(sf - kM)]\mathrm{d}M$$

yields the loop integral

$$X_k^{n,s} = \frac{1}{2\pi j}\oint\left(\frac{r}{a}\right)^n x^s z^{-k-1}\mathrm{d}z \tag{B3}$$

where the loop is the unit circle $|z| = 1$.

The integral in Eq. (B3) can be easily converted to an integral of x or y according to

$$\mathrm{d}z = \frac{r^2}{a^2\cos\phi}\frac{z}{x}\mathrm{d}x \tag{B4}$$

$$\mathrm{d}z = \frac{r}{a}\frac{z}{y}\mathrm{d}y \tag{B5}$$

L.D Wu and M.J Zhang, *Hansen Coefficients in Satellite Orbital Dynamics*,
Springer Aerospace Technology, https://doi.org/10.1007/978-981-97-0456-9

Substituting Eq. (B4) into Eq. (B3) yields the loop integral

$$X_k^{n,s} = \frac{1}{2\pi j \cos\phi} \oint \left(\frac{r}{a}\right)^{n+2} x^{s-1} z^{-k} \mathrm{d}x, \tag{B6}$$

where the loop is the unit circle $|x| = 1$.

Through the substitutions

$$\begin{aligned} &\tfrac{r}{a} = \tfrac{(1-\beta^2)^2}{1+\beta^2}(1+\beta x)^{-1}\left(1+\tfrac{\beta}{x}\right)^{-1}, \quad z = y\exp\left[-\tfrac{e}{2}(y-y^{-1})\right] \\ &y = x\tfrac{1+\beta x^{-1}}{1+\beta x}, \quad \cos\phi = \tfrac{1-\beta^2}{1+\beta^2} \end{aligned} \tag{B7}$$

we have

$$\begin{aligned} \left(\frac{r}{a}\right)^{n+2}\frac{1}{\cos\phi}x^{s-1}y^{-k} &= \frac{(1-\beta^2)^{2n+3}}{(1+\beta^2)^{n+1}}(1+\beta x)^{-n+k-2}\left(1+\beta x^{-1}\right)^{-n-k-2}x^{s-k-1} \\ \exp\left[\frac{e}{2}k(y-y^{-1})\right] &= \exp\left[\mu\beta\left(\frac{x}{1+\beta x}-\frac{x^{-1}}{1+\beta x^{-1}}\right)\right], \quad \mu = k\cos\phi \end{aligned} \tag{B8}$$

and then

$$\begin{aligned} X_k^{n,s} = &\frac{1}{2\pi j}\frac{(1-\beta^2)^{2n+3}}{(1+\beta^2)^{n+1}}\oint x^{s-k-1}(1+\beta x)^{-n+k-2}\left(1+\beta x^{-1}\right)^{-n-k-2} \\ &\times \exp\left[\mu\beta\left(\frac{x}{1+\beta x}-\frac{x^{-1}}{1+\beta x^{-1}}\right)\right]\mathrm{d}x \end{aligned} \tag{B9}$$

Hansen divided the above integrand in Eq. (B9) into the expansions of the three factors

$$\begin{aligned} &x^{s-k-1}, \quad (1+\beta x)^{-n+k-2}\exp\left[\mu\beta\left(\frac{x}{1+\beta x}\right)\right], \\ &\qquad \left(1+\beta x^{-1}\right)^{-n-k-2}\exp\left(-\frac{\mu\beta x^{-1}}{1+\beta x^{-1}}\right) \end{aligned} \tag{B10}$$

The expansion of the exponential function $\exp[\mu\beta x/(1+\beta x)]$ is

$$\exp\left[\mu\beta\left(\frac{x}{1+\beta x}\right)\right] = \sum_{m=0}^{\infty}\frac{(\mu\beta x)^m}{m!}(1+\beta x)^{-m}$$

and hence

$$(1+\beta x)^{-n+k-2}\exp\left[\mu\beta\left(\frac{x}{1+\beta x}\right)\right]=\sum_{m=0}^{\infty}\frac{(\mu\beta x)^m}{m!}(1+\beta x)^{-n+k-m-2} \quad \text{(B11)}$$

Using the binomial theorem, $(1+\beta x)^{-n+k-m-2}$ is expressed as

$$(1+\beta x)^{-n+k-m-2}=\sum_{l=0}^{\infty}(-1)^l\frac{(n-k+m+2+l-1)!}{(n-k+m+1)!l!}(\beta x)^l \quad \text{(B12)}$$

and then

$$\begin{aligned}(1+\beta x)^{-n+k-2}&\exp\left[\mu\beta\left(\frac{x}{1+\beta x}\right)\right]\\ &=\sum_{m=0}^{\infty}\frac{\mu^m}{m!}\sum_{l=0}^{\infty}(-1)^l\frac{(n-k+m+l+1)!}{(n-k+m+1)!l!}(\beta x)^{m+l}\end{aligned} \quad \text{(B13)}$$

Further defining $p=m+l$, Eq. (B13) is rewritten as

$$\begin{aligned}(1+\beta x)^{-n+k-2}&\exp\left[\mu\beta\left(\frac{x}{1+\beta x}\right)\right]\\ &=\sum_{m=0}^{\infty}\frac{\mu^m}{m!}\sum_{p=0}^{\infty}(-1)^p(-1)^m\frac{(n-k+p+1)!}{(n-k+m+1)!(p-m)!}(\beta x)^p\\ &=\sum_{p=0}^{\infty}(-1)^p\sum_{m=0}^{p}(-1)^m\frac{\mu^m}{m!}\frac{(n-k+p+1)!}{(n-k+m+1)!(p-m)!}(\beta x)^p\end{aligned} \quad \text{(B14)}$$

As $p=m+l\geq m$, the upper limit of the summation of m is p, and hence

$$(1+\beta x)^{-n+k-2}\exp\left[\mu\beta\left(\frac{x}{1+\beta x}\right)\right]=\sum_{p=0}^{\infty}(-1)^p M_p(\beta x)^p \quad \text{(B15)}$$

where

$$\begin{aligned}M_p&=\sum_{m=0}^{p}(-1)^m\frac{(n-k+p+1)!}{(n-k+m+1)!(p-m)!}\frac{\mu^m}{m!}\\ &=\sum_{m=0}^{p}(-1)^m\binom{n-k+p+1}{p-m}\frac{\mu^m}{m!}\end{aligned} \quad \text{(B16)}$$

Replacing k, μ and x in Eq. (B14) with $-k$, $-\mu$ and x^{-1} respectively (i.e., $k \Rightarrow -k$, $\mu \Rightarrow -\mu$, $x \Rightarrow x^{-1}$), the expression of $\left(1+\beta x^{-1}\right)^{-n-k-2} \exp\left[-\mu\beta\left(\frac{x^{-1}}{1+\beta x^{-1}}\right)\right]$ is immediately obtained as

$$\left(1+\beta x^{-1}\right)^{-n-k-2} \exp\left[-\mu\beta\left(\frac{x^{-1}}{1+\beta x^{-1}}\right)\right] = \sum_{q=0}^{\infty} (-1)^q N_q \beta^q x^{-q} \tag{B17}$$

where

$$N_q = \sum_{m=0}^{q} \frac{(n+k+q+1)!}{(n+k+m+1)!(q-m)!} \frac{\mu^m}{m!} = \sum_{m=0}^{q} \binom{n+k+q+1}{q-m} \frac{\mu^m}{m!} \tag{B18}$$

The product of Eqs. (B15) and (B17) is

$$\begin{aligned} &\left(1+\beta x^{-1}\right)^{-n-k-2} (1+\beta x)^{-n+k-2} \exp\left[\mu\beta\left(\frac{x}{1+\beta x} - \frac{x^{-1}}{1+\beta x^{-1}}\right)\right] \\ &= \sum_{p=0}^{\infty} \sum_{q=0}^{\infty} (-1)^{p+q} M_p N_q \beta^{p+q} x^{p-q} \end{aligned} \tag{B19}$$

Let

$$t = p - q, \quad r = p + q$$

and then $-\infty \leq t \leq \infty$. As

$$p = \frac{t+r}{2}, \quad q = \frac{r-t}{2}$$

$r \pm t$ must be even, and $r \geq |t|$. In this case, r is defined as $r = 2i + |t|$, and

$$p = \frac{t+2i+|t|}{2} = i + \frac{t+|t|}{2}, \quad q = \frac{2i+|t|-t}{2} = i + \frac{|t|-t}{2}$$

The right side of Eq. (B19) can be further expressed as

$$\sum_{t=-\infty}^{\infty} (-\beta)^{|t|} \sum_{i=0}^{\infty} M_{i+(|t|+t)/2} N_{i+(|t|-t)/2} \beta^{2i} x^t \tag{B20}$$

Multiplying Eq. (B20) by x^{s-k-1}, Eq. (B9) is rewritten as

$$X_k^{n,s} = \frac{1}{2\pi j} \frac{\left(1-\beta^2\right)^{2n+3}}{\left(1+\beta^2\right)^{n+1}} \oint \sum_{t=-\infty}^{\infty} x^{t+s-k-1} \cdot \sum_{i=0}^{\infty} M_{i+(|t|+t)/2} N_{i+(|t|-t)/2} \beta^{2i} (-\beta)^{|t|} \mathrm{d}x \tag{B21}$$

The calculation of the above loop integral is attributed to the calculation

$$\oint x^{t+s-k-1} \mathrm{d}x \tag{B22}$$

According to the theory of the complex integral, we have

$$\oint x^{t+s-k-1} \mathrm{d}x = \begin{cases} 2\pi j, & (t+s-k-1=-1) \\ 0, & (t+s-k-1 \neq -1) \end{cases} \tag{B23}$$

and then

$$t = k - s \tag{B24}$$

Substituting Eqs. (B23) and (B24) into Eq. (B21), the expression for Hansen coefficients is simplified as

$$X_k^{n,s} = \frac{\left(1-\beta^2\right)^{2n+3}}{\left(1+\beta^2\right)^{n+1}} (-\beta)^{|k-s|} \sum_{i=0}^{\infty} M_{i+(|k-s|+k-s)/2} N_{i+[|k-s|-(k-s)]/2} \beta^{2i} \tag{B25}$$

where the calculations of $M_{i+(|k-s|+k-s)/2}$ and $N_{i+[|k-s|-(k-s)]/2}$ refer to the definition expressions Eqs. (B16) and (B18) respectively. Equation (3.36) is thus proven.

M_p and N_q in Eqs. (B16) and (B18) can be expressed with the generalized Laguerre polynomials $L_p^{(n-k+1)}(\mu)$ and $L_q^{(n+k+1)}(-\mu)$, and

$$\begin{aligned} M_p &= \sum_{m=0}^{p} (-1)^m \binom{n-k+p+1}{p-m} \frac{\mu^m}{m!} = L_p^{(n-k+1)}(\mu) \\ N_q &= \sum_{m=0}^{q} \binom{n+k+q+1}{q-m} \frac{\mu^m}{m!} = L_q^{(n+k+1)}(-\mu) \end{aligned} \tag{B26}$$

The definition of the generalized Laguerre polynomials involved in Eq. (B26) is

$$L_m^{(j)}(x) = \sum_{k=0}^{m} (-1)^k \binom{m+j}{m-k} \frac{x^k}{k!} \tag{B27}$$

In addition, by substituting Eq. (B26) into Eq. (B25), Hansen coefficients are expressed as [5, 6]

$$X_k^{n,s} = \left(1-\beta^2\right)^{2n+3}\left(1+\beta^2\right)^{-(n+1)}(-\beta)^{|k-s|}\sum_{i=0}^{\infty} L_{i+a}^{(n-k+1)}(\mu)L_{i+b}^{(n+k+1)}(-\mu)\beta^{2i} \tag{B28}$$

where

$$\begin{aligned} \mu &= k\left(1-\beta^2\right)/\left(1+\beta^2\right) = k\sqrt{1-e^2} \\ a &= [|k-s| + (k-s)]/2 \\ b &= [|k-s| - (k-s)]/2 \end{aligned} \tag{B29}$$

Equation (3.38) is thus proven.

By substituting Eq. (B5) into Eq. (B3) and using the relations

$$\begin{aligned} &\frac{r}{a} = \left(1+\beta^2\right)^{-1}(1-\beta y)\left(1-\beta y^{-1}\right), \quad x = y\frac{1-\beta y^{-1}}{1-\beta y}, \\ &z = y\exp\left[-\frac{e}{2}\left(y-y^{-1}\right)\right] \end{aligned}$$

another expression of Hansen coefficients is obtained as

$$\begin{aligned} X_k^{n,s} &= \frac{1}{2\pi j}\oint y^{s-k-1}\left(1+\beta^2\right)^{-(n+1)}(1-\beta y)^{n-s+1}\left(1-\beta y^{-1}\right)^{n+s+1} \\ &\quad \exp\left[\frac{e}{2}k\left(y-y^{-1}\right)\right]\mathrm{d}y \\ &= \frac{1}{2\pi j}\oint y^{s-k-1}\left(1+\beta^2\right)^{-(n+1)}(1-\beta y)^{n-s+1}\exp(\nu\beta y)\left(1-\beta y^{-1}\right)^{n+s+1} \\ &\quad \exp\left(-\nu\beta y^{-1}\right)\mathrm{d}y \end{aligned} \tag{B30}$$

where the loop is the unit circle $|y| = 1$ and

$$\frac{e}{2}k = \nu\beta, \qquad \nu = \frac{k}{1+\beta^2} \tag{B31}$$

The integrand in Eq. (B30) contains the factors

$$(1-\beta y)^{n-s+1}\exp(\nu\beta y), \quad \left(1-\beta y^{-1}\right)^{n+s+1}\exp\left(-\nu\beta y^{-1}\right)$$

The specific expansions of these two factors are

$$\begin{aligned}\exp(\nu\beta y)(1-\beta y)^{n-s+1} &= \sum_{l=0}^{\infty}\frac{(\nu\beta y)^l}{l!}\sum_{m=0}^{\infty}(-1)^m\binom{n-s+1}{m}\beta^m y^m\\ &= \sum_{p=0}^{\infty}(-1)^p\left[\sum_{l=0}^{p}\frac{(-\nu)^l}{l!}\binom{n-s+1}{p-l}\right]\beta^p y^p\\ &= \sum_{p=0}^{\infty}(-1)^p G_p\beta^p y^p \end{aligned} \tag{B32}$$

$$\begin{aligned}\exp\left(-\nu\beta y^{-1}\right)\left(1-\beta y^{-1}\right)^{n+s+1} &= \sum_{l=0}^{\infty}\frac{\left(-\nu\beta y^{-1}\right)^l}{l!}\sum_{m=0}^{\infty}(-1)^m\binom{n+s+1}{m}\left(\beta y^{-1}\right)^m\\ &= \sum_{q=0}^{\infty}(-1)^q\sum_{l=0}^{q}\frac{\nu^l}{l!}\binom{n+s+1}{q-l}\beta^q y^{-q}\\ &= \sum_{q=0}^{\infty}(-1)^q H_q\beta^q y^{-q}\end{aligned} \tag{B33}$$

The definitions of G_p and H_q in Eqs. (B32) and (B33) are

$$G_p = \sum_{l=0}^{p}\frac{(-\nu)^l}{l!}\binom{n-s+1}{p-l} \tag{B34}$$

$$H_q = \sum_{l=0}^{q}\frac{\nu^l}{l!}\binom{n+s+1}{q-l} \tag{B35}$$

Equation (B30) then becomes

$$\begin{aligned}X_k^{n,s} &= \frac{1}{2\pi j}\oint y^{s-k-1}\left(1+\beta^2\right)^{-(n+1)}\sum_{p=0}^{\infty}(-1)^p G_p\beta^p y^p\sum_{q=0}^{\infty}(-1)^q H_q\beta^q y^{-q}\mathrm{d}y\\ &= \frac{1}{2\pi j}\oint y^{s-k-1}\left(1+\beta^2\right)^{-(n+1)}\sum_{p=0}^{\infty}G_p\sum_{q=0}^{\infty}H_q(-\beta)^{p+q}y^{p-q}\mathrm{d}y\end{aligned} \tag{B36}$$

Adopting a proof similar to that expressed by Eqs. (B25), (B36) is further rewritten as

$$X_k^{n,s} = \left(1+\beta^2\right)^{-(n+1)}(-\beta)^{|k-s|}\sum_{i=0}^{\infty}G_{i+(|k-s|+k-s)/2}H_{i+[|k-s|-(k-s)]/2}\beta^{2i} \tag{B37}$$

According to the definition of the generalized Laguerre polynomials, Eqs. (B34) and (B35) are rewritten as

$$G_p = \sum_{l=0}^{p} \frac{(-\nu)^l}{l!} \binom{n-s+1}{p-l} = L_p^{(n-s-p+1)}(\nu) \tag{B38}$$

$$H_q = \sum_{l=0}^{q} \frac{\nu^l}{l!} \binom{n+s+1}{q-l} = L_q^{(n+s-q+1)}(-\nu) \tag{B39}$$

Substituting Eqs. (B38) and (B39) into Eq. (B37), the Hansen coefficients are also expressed as [5, 6]

$$X_k^{n,s} = \left(1+\beta^2\right)^{-(n+1)} (-\beta)^{|k-s|} \sum_{i=0}^{\infty} L_{i+a}^{(n-s-i-a+1)}(\nu) L_{i+b}^{(n+s-i-b+1)}(-\nu) \beta^{2i} \tag{B40}$$

where

$$\begin{aligned} &\nu = k/\left(1+\beta^2\right), \quad a = [|k-s| + (k-s)]/2, \\ &b = [|k-s| - (k-s)]/2 \end{aligned} \tag{B41}$$

Equation (3.39) is thus proven.

Appendix C
Hill Expression of Hansen Coefficients

Defining $z = \exp(jE)$ and substituting the relations [7]

$$\exp(jf) = (1-\beta z)^{-1} z\left(1-\frac{\beta}{z}\right)$$
$$\exp(-jM) = z^{-1}\exp\left[\frac{e}{2}\left(z-\frac{1}{z}\right)\right]$$
$$\mathrm{d}M = \frac{1}{jz}\frac{r}{a}\mathrm{d}z$$
$$\frac{r}{a} = \left(1+\beta^2\right)^{-1}\left(1-\frac{\beta}{z}\right)(1-\beta z) \tag{C1}$$

into the definition of Hansen coefficients

$$X_k^{n,m} = \frac{1}{2\pi}\int_0^{2\pi}\left(\frac{r}{a}\right)^n \exp[j(mf-kM)]\mathrm{d}M \tag{C2}$$

yields the loop integral

$$\begin{aligned} X_k^{n,m} &= \frac{1}{2\pi}\oint \left(1+\beta^2\right)^{-n-1}\left(1-\frac{\beta}{z}\right)^{n+1}(1-\beta z)^{n+1} \\ &\quad \times (1-\beta z)^{-m} z^m \left(1-\frac{\beta}{z}\right)^m z^{-k}\exp\left[\frac{ke}{2}\left(z-\frac{1}{z}\right)\right]\frac{1}{jz}\mathrm{d}z \\ &= \frac{1}{2\pi j}\oint \left(1+\beta^2\right)^{-n-1}(1-\beta z)^{n-m+1}\left(1-\frac{\beta}{z}\right)^{n+m+1} \\ &\quad z^{m-k-1}\exp\left[\frac{ke}{2}\left(z-\frac{1}{z}\right)\right]\mathrm{d}z \end{aligned} \tag{C3}$$

where the loop is the unit circle $|z| = 1$.

L.D Wu and M.J Zhang, *Hansen Coefficients in Satellite Orbital Dynamics*, Springer Aerospace Technology, https://doi.org/10.1007/978-981-97-0456-9

Expanding the integrand of Eq. (C3) in a power series of z and using the relations [7]

$$\begin{aligned}
(1-\beta z)^{n-m+1} &= \sum_{s=0}^{s_1}\binom{n-m+1}{s}(-\beta)^s z^s \\
\left(1-\frac{\beta}{z}\right)^{n+m+1} &= \sum_{t=0}^{t_1}\binom{n+m+1}{t}(-\beta)^t z^{-t} \\
s_1 &= \begin{cases} n-m+1, & (n-m+1\geq 0) \\ \infty, & (n-m+1<0) \end{cases} \\
t_1 &= \begin{cases} n+m+1, & (n+m+1\geq 0) \\ \infty, & (n+m+1<0) \end{cases}
\end{aligned} \tag{C4}$$

$$J_n(x)=\frac{1}{2\pi j}\oint z^{-n-1}\exp\left[\frac{1}{2}x\left(z-\frac{1}{z}\right)\right]dz \tag{C5}$$

immediately yields an expression for Hansen coefficients:

$$\begin{aligned}
X_k^{n,m} = {} & \left(1+\beta^2\right)^{-n-1}\sum_{s=0}^{s_1}\sum_{t=0}^{t_1}\binom{n-m+1}{s}\binom{n+m+1}{t} \\
& (-\beta)^{s+t}J_{k-m-s+t}(ke)
\end{aligned} \tag{C6}$$

which is the Hill expression Eq. (3.19) presented by Giacaglia [7].

Appendix D
Fortran Programs

D.1 Fortran Programs of the Calculation Methods in Chap. 2

L.D Wu and M.J Zhang, *Hansen Coefficients in Satellite Orbital Dynamics*,
Springer Aerospace Technology, https://doi.org/10.1007/978-981-97-0456-9

!---

1) **Program Name** XNME

Calling method: CALL XNME (E, NN, XX, YY)

Function: Calculation of $X_0^{n,m}$ and their derivatives $\frac{dX_0^{n,m}}{de}$ using Eq. (2.86)

Parameters: INPUT E eccentricity

NN order

OUTPUT XX(N, M) $X_0^{n,m}$

YY(N, M) $\frac{dX_0^{n,m}}{de}$

N = 0, 1, …, NN; M = 0, 1, …, N + 1

This program calls the subroutine CJI of the calculation of binomial coefficients.

!---

```
SUBROUTINE XNME(E, NN, XX, YY)
IMPLICIT NONE
REAL*8 E
INTEGER NN
REAL*8 XX(0 : NN, 0 : NN + 1), YY(0 : NN, 0 : NN + 1)
REAL*8 EML(0 : NN + NN + NN)
INTEGER N, M
REAL*8 X, Y
INTEGER L
REAL*8 C1, C2, C3, C4, C
EML(0) = 1.0D0
DO N = 1, NN + NN + NN
    EML(N) = -E / 2.0D0 * EML(N - 1)
ENDDO
DO N = 0, NN
    DO M = 0, N + 1
        X = 0.0D0
        Y = 0.0D0
        DO L = 0, (N + 1 - M) / 2
            CALL CJI(N + 1 + M, M, C1)
            CALL CJI(L + L, L, C2)
            CALL CJI(N + 1 - M, L + L, C3)
            CALL CJI(L + M, M, C4)
            C = C1 * C2 * C3 / C4
            X = X + C * EML(M + L + L)
            Y = Y + C * DBLE(M + L + L) * EML(M + L + L) / 2.0D0 / (E / 2.0D0)
        ENDDO
        XX(N, M) = X
        YY(N, M) = Y
    ENDDO
ENDDO
RETURN
END SUBROUTINE XNME
```

```
!-----------------------------------------------------------------
    The subroutine of the calculation of binomial coefficients CJI
    SUBROUTINE CJI(J, I, C)
    IMPLICIT NONE
    INTEGER J, I
    REAL*8 C
    INTEGER K
    C = 0.0D0
    IF(I .LT. 0) RETURN
    C = 1.0D0
    IF(I .EQ. 0) RETURN
    IF(J .LT. 0) THEN
        J = -J + I - 1
        C = (-1.0D0)**I
    ENDIF
    DO K = 1, I
        C = C * DBLE(J - I + K) / DBLE(K)
    ENDDO
    RETURN
    END SUBROUTINE CJI

!-----------------------------------------------------------------
```

2) **Program Name** XNMRECURR

Calling method: CALL XNMRECURR(E, NN, XX, YY)

Function: Recursion of $X_0^{n+1,m}$ and their derivatives $\frac{dX_0^{n+1,m}}{de}$ simultaneously using Eq. (2.88)

Parameters: INPUT E eccentricity
NN order

OUTPUT XX(N + 1, M) $X_0^{n+1,m}$

YY(N + 1, M) $\frac{dX_0^{n+1,m}}{de}$

N = -1, 0, ..., NN - 1; M = 0, 1, ..., N + 2

```
!-----------------------------------------------------------------
    SUBROUTINE XNMRECURR(E, NN, XX, YY)
    IMPLICIT NONE
    REAL*8 E
    INTEGER NN
    REAL*8 XX(0 : NN, 0 : NN + 1), YY(0 : NN, 0 : NN + 1)
    INTEGER N, M
    REAL*8 Y1, Y2, X1, X2, X3
    XX(0, 0) = 1.0D0
    XX(0, 1) = -E
    YY(0, 0) = 0.0D0
    YY(0, 1) = -1.0D0
    DO N = 0, NN - 1
        DO M = 0, N + 2
            IF(M .GT. (N + 1)) THEN
                XX(N, M) = 0.0D0
                YY(N, M) = 0.0D0
```

```
            ENDIF
            Y1 = XX(N, M)
            Y2 = XX(N, IABS(M - 1))
            X1 = DBLE(N + 2)
            X2 = DBLE(N - M + 2)
            X3 = DBLE(N + M + 1)
            XX(N + 1, M) = (X2 * Y1 - X3 * E * Y2) / X1
            YY(N + 1, M) = (X2 * YY(N, M) - X3 * (E * YY(N, IABS(M - 1)) + Y2)) / X1
        ENDDO
    ENDDO
    RETURN
    END SUBROUTINE XNMRECURR
```

!---

3) **Program Name** XNMRECURR1

Calling method: CALL XNMRECURR1(E, NN, XX, YY)

Function: Recursion of the useful $X_0^{n+2,m}$ and their derivatives $\frac{dX_0^{n+2,m}}{de}$ in the perturbation calculations simultaneously using Eq. (2.90)

Parameters: INPUT E eccentricity

NN order

OUTPUT XX(N + 2, M) $X_0^{n+2,m}$

YY(N + 2, M) $\frac{dX_0^{n+2,m}}{de}$

N = -2, -1, …, NN - 2; M = MOD(N, 2), MOD(N, 2) + 2, …, N + 2

!---

```
    SUBROUTINE XNMRECURR1(E, NN, XX, YY)
    IMPLICIT NONE
    REAL*8 E
    INTEGER NN
    REAL*8 XX(0 : NN, 0 : NN + 1), YY(0 : NN, 0 : NN + 1)
    REAL*8 X
    INTEGER N, K, M
    REAL*8 Y1, Y2, X6, X7, X1, X2, X3, X4, X5
    X = 1.0D0 - E * E
    XX(0, 0) = 1.0D0
    XX(1, 1) = -1.5D0 * E
    YY(0, 0) = 0.0D0
    YY(1, 1) = -1.5D0
    DO N = 0, NN - 2
        K = MOD(N, 2)
        DO M = K, N + 2, 2
            IF(M .GT. (N + 1)) THEN
                XX(N, M) = 0.0D0
                YY(N, M) = 0.0D0
            ENDIF
            Y1 = XX(N, M)
            Y2 = XX(N + 1, IABS(M - 1))
```

```
            X6 = DBLE(N + N + 5) / DBLE(N + 3)
            X7 = DBLE(N - M + 2) * DBLE(N - M + 3) / DBLE(N + 2) / DBLE(N + 3)
            X1 = X7 * X
            X2 = X6 * E
            X3 = X1
            X4 = X2
            X5 = X7 * (E + E)
            XX(N + 2, M) = X1 * Y1 - X2 * Y2
            YY(N + 2, M) = X3 * YY(N, M) - X4 * YY(N + 1, IABS(M - 1)) &
                & - X5 * Y1 - X6 * Y2
        ENDDO
    ENDDO
    RETURN
    END SUBROUTINE XNMRECURR1

!--------------------------------------------------------
```

4) **Program Name** XRECURR

Calling method: CALL XRECURR(E, NN, XX, YY)

Function: Recursion of $X_0^{-(n+1),m}$ and their derivatives $\frac{\mathrm{d}X_0^{-(n+1),m}}{\mathrm{d}e}$ simultaneously using Eq. (2.54) and its derivative with respect to the eccentricity

Parameters: INPUT E eccentricity

NN order

OUTPUT XX(N + 1, M) $X_0^{-(n+1),m}$

YY(N + 1, M) $\frac{\mathrm{d}X_0^{-(n+1),m}}{\mathrm{d}e}$

N = 1, 2, …, NN, M = 0, 1, …, N - 1

```
!--------------------------------------------------------
    SUBROUTINE XRECURR(E, NN, XX, YY)
    IMPLICIT NONE
    REAL*8 E
    INTEGER NN
    REAL*8 XX(NN + 1, 0 : NN), YY(NN + 1, 0 : NN), X
    INTEGER N, M
    REAL*8 Y1, Y2, Y3
    X = (1.0D0 - E * E)
    XX(2, 0) = 1.0D0 / DSQRT(X)
    YY(2, 0) = E / DSQRT(X**3)
    DO N = 2, NN
        DO M = 0, N - 1
            IF(M .GE. N - 1) THEN
                XX(N, M) = 0.0D0
                YY(N, M) = 0.0D0
            ENDIF
            IF((M + 1) .GE. (N - 1)) THEN
                XX(N, M + 1) = 0.0D0
```

```
            YY(N, M + 1) = 0.0D0
        ENDIF
        Y1 = XX(N, M)
        Y2 = XX(N, M + 1)
        Y3 = XX(N, IABS(M - 1))
        XX(N + 1, M) = (Y1 + E * (Y2 + Y3) / 2.0D0) / X
        YY(N + 1, M) = (YY(N, M) + 2.0D0 * E * XX(N + 1, M) + 0.5D0 * (Y2 + Y3) &
            & + 0.5D0 * E * (YY(N, M + 1) + YY(N, IABS(M - 1)))) / X
    ENDDO
ENDDO
RETURN
END SUBROUTINE XRECURR
```

!---

5) **Program Name** YNME

Calling method: CALL YNME(E, NN, XXX, YYY)

Function: Calculation of $X_0^{-(n+1),m}$ and their derivatives $\frac{dX_0^{-(n+1),m}}{de}$ simultaneously using Eq. (2.87)

Parameters: INPUT E eccentricity

NN order

OUTPUT XX(N + 1, M) $X_0^{-(n+1),m}$

YY(N + 1, M) $\frac{dX_0^{-(n+1),m}}{de}$

N = 1, 2, …, NN; M = 0, 1, …, N - 1

This program calls the subroutine CJI of the calculation of binominal coefficients.

!---

```
SUBROUTINE YNME(E, NN, XXX, YYY)
IMPLICIT NONE
REAL*8 E
INTEGER NN
REAL*8 XXX(NN + 1, 0 : NN), YYY(NN + 1, 0 : NN)
REAL*8 EMK(0 : NN + NN + NN)
INTEGER N
REAL*8 EE, X, Y1, Y2
INTEGER M
REAL*8 A, B
INTEGER K
REAL*8 C1, C2, C
XXX = 0.0D0
YYY = 0.0D0
EMK(0) = 1.0D0
DO N = 1, NN + NN + NN
    EMK(N) = E / 2.0D0 * EMK(N - 1)
ENDDO
EE = E * E
X = 1.0D0 - EE
DO N = 0, NN
   Y1 = X**(-(N - 0.5D0))
```

```
    Y2 = (E + E) * (DBLE(N) - 0.5D0) * Y1 / X
    DO M = 0, N - 1
        A = 0.0D0
        B = 0.0D0
        DO K = 0, (N - 1 - M) / 2
            CALL CJI(N - 1, M + K + K, C1)
            CALL CJI(M + K + K, K, C2)
            C = C1 * C2
            A = A + C * EMK(M + K + K)
            B = B + C * DBLE(M + K + K) / 2.0D0 * EMK(M + K + K) / (E / 2.0D0)
        ENDDO
        XXX(N + 1, M) = Y1 * A
        YYY(N + 1, M) = Y1 * B + Y2 * A
    ENDDO
ENDDO
RETURN
END SUBROUTINE YNME
```

!---

6) **Program Name** XRECURR1

Calling method: CALL XRECURR1(E, NN, XX, YY)

Function: Calculation of $X_0^{-(n+1),m}$ and their derivatives $\dfrac{\mathrm{d}X_0^{-(n+1),m}}{\mathrm{d}e}$ simultaneously using Eq. (2.89)

Parameters: INPUT E eccentricity

NN order

OUTPUT XX(N + 1, M) $X_0^{-(n+1),m}$

YY(N + 1, M) $\dfrac{\mathrm{d}X_0^{-(n+1),m}}{\mathrm{d}e}$

N = 1, 2, …, NN; M = 0, 1, …, N - 1

!---

```
SUBROUTINE XRECURR1(E, NN, XX, YY)
IMPLICIT NONE
REAL*8 E
INTEGER NN
REAL*8 XX(NN + 1, 0 : NN), YY(NN + 1, 0 : NN), X
INTEGER N, M
REAL*8 X1, X2, Y1, Y3
X = (1.0D0 - E * E)
XX(2, 0) = 1.0D0 / DSQRT(X)
YY(2, 0) = E / DSQRT(X**3)
DO N = 2, NN
    M = N - 1
    XX(N, M) = 0.0D0
    YY(N, M) = 0.0D0
ENDDO
DO N = 2, NN
    DO M = 0, N - 1
        X1 = DBLE(N + M - 1) / DBLE(N - 1)
        X2 = X * X1
```

```
            Y1 = XX(N, M)
            Y3 = XX(N, IABS(M - 1))
            XX(N + 1, M) = (Y1 + E * Y3) / X2
            YY(N + 1, M) = (YY(N, M) + YY(N, IABS(M - 1)) * E &
                & + (E + E) * X1 * XX(N + 1, M) + Y3) / X2
        ENDDO
    ENDDO
    RETURN
    END SUBROUTINE XRECURR1
```

!--

7) **Program Name** XRECURR2

Calling method: CALL XRECURR2(E, NN, XX, YY)

Function: Recursion of the useful $X_0^{-(n+1),m}$ and their derivatives $\frac{dX_0^{-(n+1),m}}{de}$ in the perturbation calculations simultaneously using Eq. (2.91)

Parameters: INPUT E eccentricity

NN order

OUTPUT XX(N + 1, M) $X_0^{-(n+1),m}$

YY(N + 1, M) $\frac{dX_0^{-(n+1),m}}{de}$

N = 2, 3, …, NN; M = MOD(N, 2), MOD(N, 2) + 2, …, N - 2

!--

```
    SUBROUTINE XRECURR2(E, NN, XX, YY)
    IMPLICIT NONE
    REAL*8 E
    INTEGER NN
    REAL*8 XX(NN + 1, 0 : NN), YY(NN + 1, 0 : NN), X
    INTEGER N, K, M
    REAL*8 X1, X2, Y1, Y3
    X = (1.0D0 - E * E)
    XX(3, 0) = 1.0D0 / DSQRT(X**3)
    YY(3, 0) = 3.0D0 * E / DSQRT(X**5)
    DO N = 2, NN
        M = N - 1
        XX(N, M) = 0.0D0
        YY(N, M) = 0.0D0
    ENDDO
    DO N = 3, NN
        K = MOD(N, 2)
        DO M = K, N - 2, 2
            X1 = DBLE(N + M - 1) * DBLE(N + M - 2) / DBLE(N - 1)
            X2 = X * X1
            Y1 = DBLE(N - 2) * XX(N - 1, M)
            Y3 = DBLE(N + N - 3) * XX(N, IABS(M - 1))
            XX(N + 1, M) = (Y1 + E * Y3) / X2
            YY(N + 1, M) = (DBLE(N - 2) * YY(N - 1, M) &
                & + DBLE(N + N - 3) * E * YY(N, IABS(M - 1)) &
```

```
            & + (E + E) * X1 * XX(N + 1, M) + Y3) / X2
        ENDDO
    ENDDO
    RETURN
    END SUBROUTINE XRECURR2
```

!---

8) **Program Name** XNME2

Calling method: CALL XNME2(E, NN, XX, YY)

Function: Calculation of $Y_0^{n,m}$ and their derivatives $\frac{\mathrm{d}Y_0^{n,m}}{\mathrm{d}e^2}$ simultaneously using Eq. (2.95)

Parameters: INPUT E eccentricity

NN order

OUTPUT XX(N, M) $Y_0^{n,m}$

YY(N, M) $\frac{\mathrm{d}Y_0^{n,m}}{\mathrm{d}e^2}$

N = 0, 1, …, NN; M = 0, 1, …, N + 1

This program calls the subroutine CJI of the calculation of binominal coefficients.

!---

```
    SUBROUTINE XNME2(E, NN, XX, YY)
    IMPLICIT NONE
    REAL*8 E
    INTEGER NN
    REAL*8 XX(0 : NN, 0 : NN + 1), YY(0 : NN, 0 : NN + 1)
    REAL*8 HALF(0 : NN + NN)
    INTEGER N
    REAL*8 E2, E2N(0 : NN)
    INTEGER M
    REAL*8 X, Y
    INTEGER L
    REAL*8 C1, C2, C3, C4, C
    HALF(0) = 1.0D0
    DO N = 1, NN + NN
        HALF(N) = -0.5D0 * HALF(N - 1)
    ENDDO
    E2 = E * E
    E2N(0) = 1.0D0
    DO N = 1, NN
        E2N(N) = E2N(N - 1) * E2
    ENDDO
    DO N = 0, NN
        DO M = 0, N + 1
            X = 0.0D0
            Y = 0.0D0
            DO L = 0, (N + 1 - M) / 2
                CALL CJI(N + 1 + M, M, C1)
                CALL CJI(L + L, L, C2)
```

```
            CALL CJI(N + 1 - M, L + L, C3)
            CALL CJI(L + M, M, C4)
            C = C1 * C2 * C3 / C4
            X = X + C * HALF(M + L + L) * E2N(L)
            Y = Y + C * DBLE(L) * HALF(M + L + L) * E2N(L) / E2
        ENDDO
        XX(N, M) = X
        YY(N, M) = Y
    ENDDO
ENDDO
RETURN
END SUBROUTINE XNME2
```

!---

9) **Program Name** XNMRECURR12

Calling method: CALL XNMRECURR12(E, NN, XX, YY)

Function: Recursion of $Y_0^{n+1,m}$ and their derivatives $\frac{dY_0^{n+1,m}}{de^2}$ simultaneously using Eq. (2.98)

Parameters: INPUT E eccentricity

NN order

OUTPUT XX(N + 1, M) $Y_0^{n+1,m}$

YY(N + 1, M) $\frac{dY_0^{n+1,m}}{de^2}$

N = -1, 0, …, NN - 1; M = 0, 1, …, N + 2

!---

```
SUBROUTINE XNMRECURR12(E, NN, XX, YY)
IMPLICIT NONE
REAL*8 E
INTEGER NN
REAL*8 XX(0 : NN, 0 : NN + 1), YY(0 : NN, 0 : NN + 1), E2
INTEGER N, M
REAL*8 X1, X2, Y1, Y2
E2 = E * E
XX(0, 0) = 1.0D0
XX(0, 1) = -1.0D0
YY(0, 0) = 0.0D0
YY(0, 1) = 0.0D0
DO N = 0, NN - 1
    DO M = 0, N + 2
        IF(M .GT. (N + 1)) THEN
            XX(N, M) = 0.0D0
            YY(N, M) = 0.0D0
        ENDIF
        Y1 = XX(N, M)
        Y2 = XX(N, IABS(M - 1))
        IF(M .EQ. 0) THEN
            X1 = DBLE(N + 1) / DBLE(N + 2)
```

```
            XX(N + 1, M) = Y1 - X1 * E2 * Y2
            YY(N + 1, M) = YY(N, M) - X1 * (E2 * YY(N, IABS(M - 1)) + Y2)
        ELSE
            X1 = DBLE(N - M + 2) / DBLE(N + 2)
            X2 = DBLE(N + M + 1) / DBLE(N + 2)
            XX(N + 1, M) = X1 * Y1 - X2 * Y2
            YY(N + 1, M) = X1 * YY(N, M) - X2 * YY(N, IABS(M - 1))
        ENDIF
    ENDDO
ENDDO
RETURN
END SUBROUTINE XNMRECURR12
```

!---

10) **Program Name** XNMRECURR2

Calling method: CALL XNMRECURR2(E, NN, XX, YY)

Function: Recursion of the useful $Y_0^{(n+2),m}$ and their derivatives $\frac{\mathrm{d}Y_0^{(n+2),m}}{\mathrm{d}e^2}$ in the perturbation calculations simultaneously using Eq. (2.100)

Parameters: INPUT E eccentricity

NN order

OUTPUT XX(N + 2, M) $Y_0^{(n+2),m}$

YY(N + 2, M) $\frac{\mathrm{d}Y_0^{(n+2),m}}{\mathrm{d}e^2}$

N = -2, -1, …, NN - 2; M = MOD(N, 2), MOD(N, 2) + 2, …, N + 2

!---

```
SUBROUTINE XNMRECURR2(E, NN, XX, YY)
IMPLICIT NONE
REAL*8 E
INTEGER NN
REAL*8 XX(0 : NN, 0 : NN + 1), YY(0 : NN, 0 : NN + 1), E2, X
INTEGER N, K, M
REAL*8 Y1, Y2, X1, X2
E2 = E * E
X = 1.0D0 - E2
XX(0, 0) = 1.0D0
XX(1, 1) = -1.5D0
YY(0, 0) = 0.0D0
YY(1, 1) = 0.0D0
DO N = 0, NN - 2
    K = MOD(N, 2)
    DO M = K, N + 2, 2
        IF(M .GT. N + 1) THEN
            XX(N, M) = 0.0D0
            YY(N, M) = 0.0D0
```

```
        ENDIF
        Y1 = XX(N, M)
        Y2 = XX(N + 1, IABS(M - 1))
        IF(M .EQ. 0) THEN
            X1 = DBLE(N + N + 5) / DBLE(N + 3)
            XX(N + 2, M) = X * Y1 - X1 * E2 * Y2
            YY(N + 2, M) = X * YY(N, M) - Y1 &
                & - X1 * (E2 * YY(N + 1, IABS(M - 1)) + Y2)
        ELSE
            X1 = DBLE(N - M + 2) * DBLE(N - M + 3) / DBLE(N + 2) / DBLE(N + 3)
            X2 = DBLE(N + N + 5) / DBLE(N + 3)
            XX(N + 2, M) = X1 * X * Y1 - X2 * Y2
            YY(N + 2, M) = X1 * (X * YY(N, M) - Y1) - X2 * YY(N + 1, IABS(M - 1))
        ENDIF
    ENDDO
ENDDO
RETURN
END SUBROUTINE XNMRECURR2
```

!---

11) **Program Name** YRECURR

Calling method: CALL YRECURR(E, NN, XX, YY)

Function: Recursion of $Y_0^{-(n+1),m}$ and their derivatives $\frac{\mathrm{d}Y_0^{-(n+1),m}}{\mathrm{d}e^2}$ simultaneously using Eq. (7.16) in Ref. [8]

Parameters: INPUT E eccentricity

NN order

OUTPUT XX(N + 1, M) $Y_0^{-(n+1),m}$

YY(N + 1, M) $\frac{\mathrm{d}Y_0^{-(n+1),m}}{\mathrm{d}e^2}$

N = 1, 2, …, NN; M = 0, 1, …, N - 1

!---

```
SUBROUTINE YRECURR(E, NN, XX, YY)
IMPLICIT NONE
REAL*8 E
INTEGER NN
REAL*8 XX(NN + 1, 0 : NN), YY(NN + 1, 0 : NN), E2, X
INTEGER N, M
REAL*8 Y1, Y2, Y3
E2 = E * E
X = (1.0D0 - E2)
XX(2, 0) = 1.0D0 / DSQRT(X)
YY(2, 0) = 0.5D0 / DSQRT(X**3)
DO N = 2, NN
    DO M = 0, N - 1
```

```
            IF(M .GE. (N - 1)) THEN
                XX(N, M) = 0.0D0
                YY(N, M) = 0.0D0
            ENDIF
            IF((M + 1) .GE. (N - 1)) THEN
                XX(N, M + 1) = 0.0D0
                YY(N, M + 1) = 0.0D0
            ENDIF
            Y1 = XX(N, M)
            Y2 = XX(N, M + 1)
            Y3 = XX(N, IABS(M - 1))
            IF(M .EQ. 0) THEN
                XX(N + 1, M) = (Y1 + E2 * Y2) / X
                YY(N + 1, M) = (YY(N, M) + E2 * YY(N, M + 1) + Y2 + XX(N + 1, M)) / X
            ELSE
                XX(N + 1, M) = (Y1 + (E2 * Y2 + Y3) / 2.0D0) / X
                YY(N + 1, M) = (XX(N + 1, M) + YY(N, M) &
                    & + 0.5D0 * (Y2 + E2 * YY(N, M + 1) + YY(N, M - 1))) / X
            ENDIF
        ENDDO
    ENDDO
    RETURN
    END SUBROUTINE YRECURR
```

!--

12) **Program Name** YNME1

Calling method: CALL YNME1(E, NN, XXX, YYY)

Function: Calculation of $Y_0^{-(n+1),m}$ and their derivatives $\frac{dY_0^{-(n+1),m}}{de^2}$ simultaneously using Eq. (2.96)

Parameters: INPUT E eccentricity

NN order

OUTPUT XX(N + 1, M) $Y_0^{-(n+1),m}$

YY(N + 1, M) $\frac{dY_0^{-(n+1),m}}{de^2}$

N = 1, 2, …, NN; M = 0, 1, …, N − 1

This program calls the subroutine CJI of the calculation of binomial coefficients.

!--

```
    SUBROUTINE YNME1(E, NN, XXX, YYY)
    IMPLICIT NONE
    REAL*8 E
    INTEGER NN
    REAL*8 XXX(NN + 1, 0 : NN), YYY(NN + 1, 0 : NN)
    REAL*8 EE, EMK(0 : NN), HALF(0 : NN)
    INTEGER N
```

```
REAL*8 X, Y1, Y2
INTEGER M
REAL*8 A, B
INTEGER K
REAL*8 C1, C2, C
XXX = 0.0D0
YYY = 0.0D0
EE = E * E
EMK(0) = 1.0D0
HALF(0) = 1.0D0
DO N = 1, NN
    EMK(N) = EE / 4.0D0 * EMK(N - 1)
    HALF(N) = 0.5D0 * HALF(N - 1)
ENDDO
X = 1.0D0 - EE
DO N = 1, NN
    Y1 = X**(-(N - 0.5D0))
    Y2 = (DBLE(N) - 0.5D0) * Y1 / X
    DO M = 0, N - 1
        A = 0.0D0
        B = 0.0D0
        DO K = 0, (N - 1 - M) / 2
            CALL CJI(N - 1, M + K + K, C1)
            CALL CJI(M + K + K, K, C2)
            C = C1 * C2
            A = A + C * EMK(K)
            IF(K .NE. 0) B = B + C * DBLE(K) / 4.0D0 * EMK(K - 1)
        ENDDO
        XXX(N + 1, M) = Y1 * A * HALF(M)
        YYY(N + 1, M) = (Y1 * B + Y2 * A) * HALF(M)
    ENDDO
ENDDO
RETURN
END SUBROUTINE YNME1
```

```
!---------------------------------------------------
```

13) **Program Name** YRECURR1

Calling method: CALL YRECURR1(E, NN, XX, YY)

Function: Recursion of $Y_0^{-(n+1),m}$ and their derivatives $\frac{\mathrm{d}Y_0^{-(n+1),m}}{\mathrm{d}e^2}$ simultaneously using Eq. (2.99)

Parameters: INPUT E eccentricity

NN order

OUTPUT XX(N + 1, M) $Y_0^{-(n+1),m}$

YY(N + 1, M) $\frac{dY_0^{-(n+1),m}}{de^2}$

N = 1, 2, …, NN; M = 0, 1, …, N - 1

```
!-----------------------------------------------------
      SUBROUTINE YRECURR1(E, NN, XX, YY)
      IMPLICIT NONE
      REAL*8 E
      INTEGER NN
      REAL*8 XX(NN + 1, 0 : NN), YY(NN + 1, 0 : NN), E2, X
      INTEGER N, M
      REAL*8 Y1, Y3, X1
      E2 = E * E
      X = (1.0D0 - E2)
      XX(2, 0) = 1.0D0 / DSQRT(X)
      YY(2, 0) = 0.5D0 / DSQRT(X**3)
      DO N = 2, NN
          M = N - 1
          XX(N, M) = 0.0D0
          YY(N, M) = 0.0D0
      ENDDO
      DO N = 2, NN
          DO M = 0, N - 1
              Y1 = XX(N, M)
              Y3 = XX(N, IABS(M - 1))
              X1 = DBLE(N - 1) / DBLE(N + M - 1)
              IF(M .EQ. 0) THEN
                  XX(N + 1, M) = (Y1 + E2 * Y3) / X
                  YY(N + 1, M) = (YY(N, M) + E2 * YY(N, IABS(M - 1)) + Y3 &
                      &   XX(N + 1, M)) / X
              ELSE
                  XX(N + 1, M) = (Y1 + Y3) * X1 / X
                  YY(N + 1, M) = ((YY(N, M) + YY(N, M - 1)) * X1 + XX(N + 1, M)) / X
              ENDIF
          ENDDO
      ENDDO
      RETURN
      END SUBROUTINE YRECURR1

!-----------------------------------------------------
```

14) **Program Name** YRECURR2

Calling method: CALL YRECURR2(E, NN, XX, YY)

Function: Recursion of the useful $Y_0^{-(n+1),m}$ and their derivatives $\frac{dY_0^{-(n+1),m}}{de^2}$ in the perturbation calculations simultaneously using Eq. (2.101)

Parameters: INPUT E eccentricity

NN order

OUTPUT XX(N + 1, M) $Y_0^{-(n+1),m}$

YY(N + 1, M) $\frac{dY_0^{-(n+1),m}}{de^2}$

N = 2, 3, …, NN; M = MOD(N, 2), MOD(N, 2) + 2, …, N - 2

```
!-------------------------------------------------------
      SUBROUTINE YRECURR2(E, NN, XX, YY)
      IMPLICIT NONE
      REAL*8 E
      INTEGER NN
      REAL*8 XX(NN + 1, 0 : NN ), YY(NN + 1, 0 : NN), E2, X
      INTEGER N, K, M
      REAL*8 Y1, Y3, X1
      E2 = E * E
      X = (1.0D0 - E2)
      XX(3, 0) = 1.0D0 / DSQRT(X**3)
      YY(3, 0) = 1.5D0 / DSQRT(X**5)
      DO N = 2, NN
          M = N -1
          XX(N, M) = 0.0D0
          YY(N, M) = 0.0D0
      ENDDO
      DO N = 3, NN
          K = MOD(N, 2)
          DO M = K, N - 2, 2
              Y1 = DBLE(N + N - 3) * XX(N, IABS(M - 1))
              Y3 = DBLE(N - 2) * XX(N - 1, M)
              X1 = DBLE(N - 1) / DBLE(N + M - 1) / DBLE(N + M - 2)
              IF(M .EQ. 0) THEN
                  XX(N + 1, M) = (Y1 * E2 + Y3) / DBLE(N - 2) / X
                  YY(N + 1, M) = (DBLE(N + N - 3) / DBLE(N - 2) &
                      & * (E2 * YY(N, IABS(M - 1)) + XX(N, IABS(M - 1))) &
                      & + YY(N - 1, M) + XX(N + 1, M)) / X
              ELSE
                  XX(N + 1, M) = X1 * (Y1 + Y3) / X
                  YY(N + 1, M) = ((DBLE(N + N - 3) * YY(N, M - 1) &
                      & + DBLE(N - 2) * YY(N - 1, M)) * X1 + XX(N + 1, M)) / X
              ENDIF
          ENDDO
      ENDDO
      RETURN
      END SUBROUTINE YRECURR2

!-------------------------------------------------------
```

15) **Program Name** ZRECURR
Calling method: CALL ZRECURR(E, NN, XX, YY)

Function: Calculation of $Y_0^{-(n+1),m}$ and their derivatives $\frac{\mathrm{d}Y_0^{-(n+1),m}}{\mathrm{d}e^2}$ simultaneously using Eqs. (2.103) and (2.105)

Parameters: INPUT E eccentricity

NN order

OUTPUT XX(N + 1, M) $Y_0^{-(n+1),m}$

YY(N + 1, M) $\frac{\mathrm{d}Y_0^{-(n+1),m}}{\mathrm{d}e^2}$

N = 1, 2, …, NN; M = 0, 1, …, N - 1

```
!---------------------------------------------------------
      SUBROUTINE ZRECURR(E, NN, XX, YY)
      IMPLICIT NONE
      REAL*8 E
      INTEGER NN
      REAL*8 XX(NN + 1, 0 : NN), YY(NN + 1, 0 : NN), E2, X
      INTEGER N, M
      REAL*8 X1, Y1, Y3, X2
      E2 = E * E
      X = (1.0D0 - E2)
      XX(2, 0) = 1.0D0
      YY(2, 0) = 0.0D0
      DO N = 2, NN
          M = N - 1
          XX(N, M) = 0.0D0
          YY(N, M) = 0.0D0
      ENDDO
      DO N = 2, NN
          DO M = 0, N - 1
              X1 = DBLE(N + M - 1) / DBLE(N - 1)
              Y1 = XX(N, M)
              Y3 = XX(N, IABS(M - 1))
              IF(M .EQ. 0) THEN
                  XX(N + 1, M) = Y1 + E2 * Y3
                  YY(N + 1, M) = YY(N, M) + E2 * YY(N, IABS(M - 1)) + Y3
              ELSE
                  XX(N + 1, M) = (Y1 + Y3) / X1
                  YY(N + 1, M) = (YY(N, M) + YY(N, IABS(M - 1))) / X1
              ENDIF
          ENDDO
      ENDDO
      DO N = 1, NN
          X1 = X**(-(N - 0.5D0))
          X2 = X1 / X
          DO M = 0, N - 1
              YY(N + 1, M) = X1 * YY(N + 1, M) + X2 * (DBLE(N) - 0.5D0) * XX(N + 1, M)
              XX(N + 1, M) = X1 * XX(N + 1, M)
```

```
    ENDDO
ENDDO
RETURN
END SUBROUTINE ZRECURR
```

!---

16) **Program Name** ZRECURR1

Calling method: CALL ZRECURR1(E, NN, XX, YY)

Function: Recursion of the useful $Y_0^{-(n+1),m}$ and their derivatives $\frac{\mathrm{d}Y_0^{-(n+1),m}}{\mathrm{d}e^2}$ in the perturbation calculations simultaneously using Eqs. (2.104) and (2.105)

Parameters: INPUT E eccentricity

NN order

OUTPUT XX(N + 1, M) $Y_0^{-(n+1),m}$

YY(N + 1, M) $\frac{\mathrm{d}Y_0^{-(n+1),m}}{\mathrm{d}e^2}$

N = 2, 3, …, NN; M = MOD(N, 2), MOD(N, 2) + 2, …, N - 2

!---

```
SUBROUTINE ZRECURR1(E, NN, XX, YY)
IMPLICIT NONE
REAL*8 E
INTEGER NN
REAL*8 XX(NN + 1, 0 : NN), YY(NN + 1, 0 : NN), E2, X
INTEGER N, K, M
REAL*8 X1, X2, X3, X4, Y1, Y3
E2 = E * E
X = (1.0D0 - E2)
XX(3, 0) = 1.0D0
YY(3, 0) = 0.0D0
DO N = 2, NN
    M = N - 1
    XX(N, M) = 0.0D0
    YY(N, M) = 0.0D0
ENDDO
DO N = 3, NN
    K = MOD(N, 2)
    DO M = K, N - 2, 2
        X1 = DBLE(N + N - 3) / DBLE(N - 2)
        X2 = X1 * E2
        X3 = DBLE(N - 1) / DBLE(N + M - 1) / DBLE(N + M - 2)
        X4 = DBLE(N - 2) * X
        Y1 = XX(N - 1, M)
        Y3 = XX(N, IABS(M - 1))
        IF(M .EQ. 0) THEN
            XX(N + 1, M) = X * Y1 + X2 * Y3
```

```
            YY(N + 1, M) = X * YY(N - 1, M) &
                & - Y1 + X1 * (E2 * YY(N, IABS(M - 1)) + Y3)
        ELSE
            XX(N + 1, M) = X3 * (DBLE(N + N - 3) * Y3 + X4 * Y1)
            YY(N + 1, M) = X3 * (DBLE(N + N - 3) * YY(N, IABS(M - 1)) &
                & + X4 * YY(N - 1, M) - DBLE(N - 2) * Y1)
        ENDIF
    ENDDO
ENDDO
DO N = 2, NN
    X1 = X**(-(N - 0.5D0))
    X2 = X1 / X
    K = MOD(N, 2)
    DO M = K, N - 2, 2
        YY(N + 1, M) = X1 * YY(N + 1, M) &
            & + X2 * (DBLE(N) - 0.5D0) * XX(N + 1, M)
        XX(N + 1, M) = X1 * XX(N + 1, M)
    ENDDO
ENDDO
RETURN
END SUBROUTINE ZRECURR1
```

D.2 Fortran Programs of the Calculation Methods in Chap. 3

```
!-----------------------------------------------------------
```

1) **Program Name** WNUK2

Calling method: CALL WNUK2(N, M, K, E, X, DX)

Function: Calculation of Hansen coefficients $X_k^{n,m}$ and their derivatives $\frac{dX_k^{n,m}}{de}$ using Eqs. (3.87) and (3.74b)

Parameters: INPUT E eccentricity

N, M, K order and degree

OUTPUT X $X_k^{n,m}$

DX $\frac{dX_k^{n,m}}{de}$

This program calls the subroutine WNUK1.

```
!-----------------------------------------------------------
SUBROUTINE WNUK2(N, M, K, E, X, DX)
IMPLICIT NONE
INTEGER N, M, K
REAL*8 E, X, DX
REAL*8 E1, XNMK1, XNMK2
E1 = 1.0D0 - E * E
CALL WNUK1(N, M, K, E, X)
CALL WNUK1(N, M - 1, K, E, XNMK1)
```

```
CALL WNUK1(N, M - 2, K, E, XNMK2)
DX = (-(DBLE(2 * M) + DBLE((N + M)) * E * E &
    & - DBLE(2 * K) * E1 * DSQRT(E1)) * X &
    & - DBLE((2 * N + 4 * M)) * E * XNMK1 &
    & - DBLE((N + M)) * E * E * XNMK2) / 2.0D0 / E / E1
RETURN
END SUBROUTINE WNUK2
```

!--

2) **Program Name** WNUK

Calling method: CALL WNUK(N, M, K, E, X, DX)

Function: Calculation of Hansen coefficients $X_k^{n,m}$ and their derivatives $\frac{dX_k^{n,m}}{de}$ using Eqs. (3.87) and (3.88)

Parameters: INPUT E eccentricity

N, M, K indexes

OUTPUT X $X_k^{n,m}$

DX $\frac{dX_k^{n,m}}{de}$

This program calls the subroutines BESSEL and CJI.

!--

```
SUBROUTINE WNUK(N, M, K, E, X, DX)
IMPLICIT NONE
INTEGER N, M, K
REAL*8 E, X, DX
REAL*8 BETA, B2, DBDE, JN(-71 : 71)
INTEGER T, Q, S
REAL*8 C1, C2, C, DCDB, DCDE, DJDE
BETA = E / (1.0D0 + DSQRT(1.0D0 - E * E))
B2 = BETA * BETA
DBDE = (1.0D0 + B2)**2 / (1.0D0 - B2) / 2.0D0
CALL BESSEL(71, DBLE(K) * E, JN)
X = 0.0D0
DX = 0.0D0
DO T = -70, 70
    Q = K - T - M
    DO S = 0, 70
        IF(Q .GE. 0) THEN
            CALL CJI(N - M + 1, Q + S, C1)
            CALL CJI(N + M + 1, S, C2)
        ELSE
            CALL CJI(N + M + 1, -Q + S, C1)
            CALL CJI(N - M + 1, S, C2)
        ENDIF
        C = (-1.0D0)**ABS(Q) * C1 * C2 * (1.0D0 + B2)**(-(N + 1)) &
```

```
            & * BETA**(S + S + ABS(Q))
        DCDB = C * (-2.0D0 * BETA * (DBLE(N) + 1.0D0) / (1.0D0 + B2) &
            & + (DBLE(S) + DBLE(S) + DBLE(ABS(Q))) / BETA)
        DCDE = DCDB * DBDE
        DJDE = (JN(T - 1) - JN(T + 1)) / 2.0D0
        X = X + C * JN(T)
        DX = DX + DCDE * JN(T) + C * DBLE(K) * DJDE
    ENDDO
ENDDO
RETURN
END SUBROUTINE WNUK
```

!--

The subroutine BESSEL
Calling method: CALL BESSEL(N, X, JN)
Function: Calculation of Bessel functions
Parameters: INPUT X parameter
N order
OUTPUT JN(-N : N) Bessel functions

!--

```
SUBROUTINE BESSEL(N, X, JN)
IMPLICIT NONE
INTEGER N
REAL*8 X, JN(-N : N)
REAL*8 PI, X2, X3, X4, A, B, Y
INTEGER K
REAL*8 P(N + N + 1)
INTEGER S
PI = 3.14159265358979323846264 3D0
X2 = X * X
IF(DABS(X) .GE. 20.0D0) THEN
    X3 = X * X2
    X4 = X2 * X2
    A = 1.0D0 - 0.0625D0 / X2 + 53.0D0 / 512.0D0 / X4
    B = -0.125D0 / DABS(X) + 25.0D0 / 384.0D0 / DABS(X3)
    JN(0) = DSQRT(2.0D0 / PI / DABS(X)) * A * DCOS(DABS(X) - PI / 4.0D0 + B)
ELSE
    A = 1.0D0
    Y = 0.0D0
    K = 1
    DO WHILE(DABS(A) .GE. 1.0D-15)
        Y = Y + A
        A = A * (-0.25D0 * X2) / DBLE(K) / DBLE(K)
        K = K + 1
    ENDDO
    JN(0) = Y
ENDIF
P(N + N + 1) = 0.0D0
```

```
DO S = N + N, 1, -1
    IF(X .EQ. 0.0D0) THEN
        P(S) = 0.0D0
    ELSE
        P(S) = 1.0D0 / (DBLE(S + S) / X - P(S + 1))
    ENDIF
ENDDO
DO S = 1, N
    JN(S) = JN(S - 1) * P(S)
    JN(-S) = (-1.0D0)**S * JN(S)
ENDDO
RETURN
END SUBROUTINE BESSEL
```

!--

The subroutine CJI

Calling method: CALL CJI(J, I, C)

Function: Calculation of the binomial coefficient $\begin{pmatrix} j \\ i \end{pmatrix}$

Parameters: INPUT J, I parameters

OUTPUT C $\begin{pmatrix} j \\ i \end{pmatrix}$

!--

```
SUBROUTINE CJI(J, I, C)
IMPLICIT NONE
INTEGER J, I
REAL*8 C
INTEGER K
C = 0.0D0
IF(I .LT. 0) RETURN
C = 1.0D0
IF(I .EQ. 0) RETURN
IF(J .LT. 0) THEN
    J = -J + I - 1
    C = (-1.0D0)**I
ENDIF
DO K = 1, I
    C = C * DBLE(J - I + K) / DBLE(K)
ENDDO
RETURN
END SUBROUTINE CJI
```

!--

3) **Program Name** WNUK1

Calling method: CALL WNUK1(N, M, K, E, X)

Function: Calculation of Hansen coefficients $X_k^{n,m}$ using Eq. (3.87)

Parameters: INPUT E eccentricity

N, M, K indexes

OUTPUT X $X_k^{n,m}$

This program calls the subroutines BESSEL and CJI.

```
!-----------------------------------------------------------
      SUBROUTINE WNUK1(N, M, K, E, X)
      IMPLICIT NONE
      INTEGER N, M, K
      REAL*8 E, X
      REAL*8 BETA, B2, JN(-70 : 70)
      INTEGER T, Q, S
      REAL*8 C1, C2, C
      BETA = E / (1.0D0 + DSQRT(1.0D0 - E * E))
      B2 = BETA * BETA
      CALL BESSEL(70, DBLE(K) * E, JN)
      X = 0.0D0
      DO T = -70, 70
          Q = K - T - M
          DO S = 0, 70
              IF(Q .GE. 0) THEN
                  CALL CJI(N - M + 1, Q + S, C1)
                  CALL CJI(N + M + 1, S, C2)
              ELSE
                  CALL CJI(N + M + 1, -Q + S, C1)
                  CALL CJI(N - M + 1, S, C2)
              ENDIF
              C = (-1.0D0)**ABS(Q) * C1 * C2 * (1.0D0 + B2)**(-(N + 1)) &
                  & * BETA**(S + S + ABS(Q))
              X = X + C * JN(T)
          ENDDO
      ENDDO
      RETURN
      END SUBROUTINE WNUK1

!-----------------------------------------------------------
```

4) **Program Name** MCCLAIN2

Calling method: CALL MCCLAIN2 (N, S, T, E, X, DXDE)

Function: Calculation of Hansen coefficients $X_t^{n,s}$ and their derivatives $\frac{dX_t^{n,s}}{de}$ using Eq. (3.39)

Parameters: INPUT E eccentricity

N, S, T order and degree

OUTPUT X $X_t^{n,s}$

DXDE $\frac{dX_t^{n,s}}{de}$

This program calls the subroutine LAGUERRE.

```
!-------------------------------------------------------------
    SUBROUTINE MCCLAIN2(N, S, T, E, XX, DXDE)
    IMPLICIT NONE
    INTEGER N, S, T
    REAL*8 E, XX, DXDE
    REAL*8 BETA, B2, X, DXDB2, AA, DADB2
    INTEGER NN, NNN, A, B
    REAL*8 KX, DX
    INTEGER I, ALPHA
    REAL*8 LAQUA(-1 : 200), DLAQUA(-1 : 200), LAQUB(-1 : 200), DLAQUB(-1 : 200)
    REAL*8 BETA2I, DLDX, HALF, ETS, DKDE2, DKDE
    BETA = E / (1.0D0 + DSQRT(1.0D0 - E * E))
    B2 = BETA * BETA
    X = DBLE(T) / (1.0D0 + B2)
    DXDB2 = -DBLE(T) / (1.0D0 + B2) / (1.0D0 + B2)
    AA = (1.0D0 + B2)**(-(N + 1) + ABS(T - S))
    DADB2 = DBLE(-(N + 1) + ABS(T - S)) * AA / (1.0D0 + B2)
    NN = 70
    NNN = NN + ABS(T - S)
    A = (ABS(T - S) + (T - S)) / 2
    B = (ABS(T - S) - (T - S)) / 2
    KX = 0.0D0
    DX = 0.0D0
    DO I = NN, 0, -1
        ALPHA = N - S - I - A + 1
        CALL LAGUERRE(NNN, ALPHA, X, LAQUA)
        CALL LAGUERRE(NNN, ALPHA + 1, X, DLAQUA)
        ALPHA = N + S - I - B + 1
        CALL LAGUERRE(NNN, ALPHA, -X, LAQUB)
        CALL LAGUERRE(NNN, ALPHA + 1, -X, DLAQUB)
        BETA2I = BETA**(I + I)
        KX = KX + LAQUA(I + A) * LAQUB(I + B) * BETA2I
        DLDX = LAQUA(I + A) * DLAQUB(I + B - 1) - DLAQUA(I + A - 1) * LAQUB(I + B)
        DX = DX + DLDX * DXDB2 * BETA2I + DBLE(I) * LAQUA(I + A) * LAQUB(I + B) &
            & * BETA**(I + I - 2)
    ENDDO
    HALF = (-0.5D0)**ABS(T - S)
    ETS = E**ABS(T - S)
    XX = KX * AA * HALF * ETS
    DKDE2 = (KX * DADB2 + DX * AA) * HALF * ETS * (1.0D0 + B2)**3 / (1.0D0 - B2) / 4.0D0
    DKDE = (E + E) * DKDE2
    DXDE = DKDE + KX * AA * HALF * DBLE(ABS(T - S)) * E**(ABS(T - S) - 1)
    RETURN
    END SUBROUTINE MCCLAIN2
```

!---

The subroutine LAGUERRE

Calling method: CALL LAGUERRE (N, A, X, LAQU)

Function: Calculation of Laguerre polynomials $L_m^{(a)}(x)$ using recursion formula Eq. (3.54)

Parameters: INPUT X independent variable

A parameter

N upper limit of m

OUTPUT LAQUE(-1 : N) $L_m^{(a)}(x)$, $(m = -1, 0, 1, \cdots, N)$

!---

```
SUBROUTINE LAGUERRE(N, A, X, LAQU)
IMPLICIT NONE
INTEGER N, A
REAL*8 X, LAQU(-1 : N)
INTEGER I
LAQU(-1) = 0.0D0
LAQU(0) = 1.0D0
LAQU(1) = DBLE(A) + 1.0D0 - X
DO I = 1, N - 1
    LAQU(I + 1) = ((DBLE(I) + 1.0D0 - X) * LAQU(I) &
        & + DBLE(I + A) * (LAQU(I) - LAQU(I - 1))) / DBLE(I + 1)
ENDDO
RETURN
END SUBROUTINE LAGUERRE
```

!---

5) **Program Name** MCCLAIN1

Calling method: CALL MCCLAIN1 (N, S, T, E, X, DXDE)

Function: Calculation of Hansen coefficients $X_t^{n,s}$ and their derivatives $\frac{dX_t^{n,s}}{de}$ using Eqs. (3.76) and (3.79)

Parameters: INPUT E eccentricity

N, S, T indexes

OUTPUT X $X_t^{n,s}$

DXDE $\frac{dX_t^{n,s}}{de}$

This program calls the subroutine LAGUERRE.

!---

```
SUBROUTINE MCCLAIN1(N, S, T, E, XX, DXDE)
IMPLICIT NONE
INTEGER N, S, T
REAL*8 E, XX, DXDE
REAL*8 BETA, B2, X, AA, DADB2, DXDB2
```

```
INTEGER NN, NNN, A, B
REAL*8 KX, DX
INTEGER ALPHA
REAL*8 LAQUA(-1 : 200), DLAQUA(-1 : 200), LAQUB(-1 : 200), DLAQUB(-1 : 200)
INTEGER I
REAL*8 BETA2I, DLDX, HALF, ETS, DKDE2, DKDE
BETA = E / (1.0D0 + DSQRT(1.0D0 - E * E))
B2 = BETA * BETA
X = DBLE(T) * (1.0D0 - B2) / (1.0D0 + B2)
AA = (1.0D0 - B2)**(2 * N + 3) * (1.0D0 + B2)**(-(N + 1) + ABS(T - S))
DADB2 = AA * (-DBLE(2 * N + 3) / (1.0D0 - B2) + DBLE(-(N + 1) &
    & + ABS(T - S)) / (1.0D0 + B2))
DXDB2 = -2.0D0 * DBLE(T) / (1.0D0 + B2) / (1.0D0 + B2)
NN = 70
NNN = NN + ABS(T - S)
A = (ABS(T - S) + (T - S)) / 2
B = (ABS(T - S) - (T - S)) / 2
KX = 0.0D0
DX = 0.0D0
ALPHA = N - T + 1
CALL LAGUERRE(NNN, ALPHA, X, LAQUA)
CALL LAGUERRE(NNN, ALPHA + 1, X, DLAQUA)
ALPHA = N + T + 1
CALL LAGUERRE(NNN, ALPHA, -X, LAQUB)
CALL LAGUERRE(NNN, ALPHA + 1, -X, DLAQUB)
DO I = NN, 0, -1
    BETA2I = BETA**(I + I)
    KX = KX + LAQUA(I + A) * LAQUB(I + B) * BETA2I
    DLDX = LAQUA(I + A) * DLAQUB(I + B - 1) - DLAQUA(I + A - 1) * LAQUB(I + B)
    DX = DX + DLDX * DXDB2 * BETA2I + DBLE(I) * LAQUA(I + A) * LAQUB(I + B) &
        & * BETA**(I + I - 2)
ENDDO
HALF = (-0.5D0)**ABS(T - S)
ETS = E**ABS(T - S)
XX = KX * AA * HALF * ETS
DKDE2 =(KX * DADB2 + DX * AA) * HALF * ETS * (1.0D0 + B2)**3 / (1.0D0 - B2) / 4.0D0
DKDE = (E + E) * DKDE2
DXDE = DKDE + DBLE(ABS(T - S)) * KX * AA * HALF * E**(ABS(T - S) - 1)
RETURN
END SUBROUTINE MCCLAIN1
```

```
!-----------------------------------------------------------------
```

6) **Program Name** GIACAGLIA

Calling method: CALL GIACAGLIA(L, P, Q, E, X, DX)

Function: Calculation of eccentricity functions G_{lpq} and their derivatives $\frac{dG_{lpq}}{de}$ using Eqs. (3.90) and (3.91)

Parameters: INPUT E eccentricity
L, P, Q indexes
OUTPUT X G_{lpq}
DX $\frac{dG_{lpq}}{de}$

This program calls the subroutines BESSEL and CJI.

```
!--------------------------------------------------------
    SUBROUTINE GIACAGLIA(L, P, Q, E, X, DX)
    IMPLICIT NONE
    INTEGER L, P, Q
    REAL*8 E, X, DX
    REAL*8 BETA, B2, A, DBDE
    INTEGER K
    REAL*8 JN(-200 : 200)
    INTEGER S, T
    REAL*8 C1, C2, DJDE, C, DCDE
    BETA = E / (1.0D0 + DSQRT(1.0D0 - E * E))
    B2 = BETA * BETA
    A = (1.0D0 + B2)**L
    DBDE = (1.0D0 + B2)**2 / (1.0D0 - B2) / 2.0D0
    K = L - P - P + Q
    CALL BESSEL(200, DBLE(K) * E, JN)
    X = 0.0D0
    DX = 0.0D0
    IF(P .NE. 0 .AND. P .NE. L) THEN
        DO S = 0, 70
            DO T = 0, 70
                CALL CJI(L + L - P - P + S - 1, S, C1)
                CALL CJI(P + P + T - 1, T, C2)
                DJDE = DBLE(K) * (JN(Q - S + T - 1) - JN(Q - S + T + 1)) / 2.0D0
                C = C1 * C2 * A * BETA**(S + T)
                DCDE = (DBLE(L + L) * BETA / (1.0D0 + B2) &
                    & + DBLE(S + T) / BETA) * C * DBDE
                X = X + C * JN(Q - S + T)
                DX = DX + DCDE * JN(Q - S + T) + C * DJDE
            ENDDO
        ENDDO
    ENDIF
    IF(P .EQ. L .AND. P .NE. 0) THEN
        S = 0
        DO T = 0, 70
            CALL CJI(L + L - P - P + S - 1, S, C1)
            CALL CJI(P + P + T - 1, T, C2)
            DJDE = DBLE(K) * (JN(Q - S + T - 1) - JN(Q - S + T + 1)) / 2.0D0
            C = C1 * C2 * A * BETA**(S + T)
            DCDE = (DBLE(L + L) * BETA / (1.0D0 + B2) + DBLE(S + T) / BETA) * C * DBDE
            X = X + C * JN(Q - S + T)
```

```
            DX = DX + DCDE * JN(Q - S + T) + C * DJDE
        ENDDO
    ENDIF
    IF(P .EQ. 0 .AND. P .NE. L) THEN
        DO S = 0, 70
            T = 0
            CALL CJI(L + L - P - P + S - 1, S, C1)
            CALL CJI(P + P + T - 1, T, C2)
            DJDE = DBLE(K) * (JN(Q - S + T - 1) - JN(Q - S + T + 1)) / 2.0D0
            C = C1 * C2 * A * BETA**(S + T)
            DCDE = (DBLE(L + L) * BETA / (1.0D0 + B2) + DBLE(S + T) / BETA) * C * DBDE
            X = X + C * JN(Q - S + T)
            DX = DX + DCDE * JN(Q - S + T) + C * DJDE
        ENDDO
    ENDIF
    RETURN
    END SUBROUTINE GIACAGLIA
```

```
!-----------------------------------------------------------
```

7) **Program Name** BALMINO

Calling method: CALL BALMINO(N, M1, S1, E, X, DX)

Function: Calculation of Hansen coefficients $X_{m+s}^{n,m}$ and their derivatives $\frac{dX_{m+s}^{n,m}}{de}$ using Eqs. (3.10) and (3.86)

Parameters: INPUT E eccentricity

N, M1, S1 indexes

OUTPUT X $X_{m+s}^{n,m}$

DX $\frac{dX_{m+s}^{n,m}}{de}$

This program calls the subroutine CJI.

```
!-----------------------------------------------------------
    SUBROUTINE BALMINO(N, M1, S1, E, X, DX)
    IMPLICIT NONE
    INTEGER N, M1, S1
    REAL*8 E, X, DX
    REAL*8 HALFE
    INTEGER M, S, T
    REAL*8 XST
    INTEGER J
    REAL*8 XP
    INTEGER P
    REAL*8 AP, XQ
    INTEGER Q
    REAL*8 BQ
    INTEGER K
    REAL*8 C1, C2
```

```
HALFE = E / 2.0D0
M = M1
S = S1
IF(S1 .LT. 0) THEN
    M = -M1
    S = -S1
ENDIF
X = 0.0D0
DX = 0.0D0
DO T = 0, 50
    XST = 0.0D0
    DO J = 0, T
        XP = 1.0D0
        DO P = 0, J
            CALL CJI(N + M + 1, J - P, AP)
            IF(P .NE. 0) XP = XP * DBLE(S + M) / DBLE(P)
            XQ = 1.0D0
            DO Q = 0, S + J
                CALL CJI(N - M + 1, S + J - Q, BQ)
                IF(Q .NE. 0) XQ = -XQ * DBLE(S + M) / DBLE(Q)
                K = T + T - N + S - P - Q - 2
                CALL CJI(K, T - J, C1)
                CALL CJI(K, T - J - 1, C2)
                XST = XST + AP * BQ * XP * XQ * (C1 - C2)
            ENDDO
        ENDDO
    ENDDO
    X = X + XST * HALFE**(T + T + S)
    DX = DX + XST * DBLE(T + T + S) * HALFE**(T + T + S - 1) / 2.0D0
ENDDO
X = X * (-1.0D0)**S
DX = DX * (-1.0D0)**S
RETURN
END SUBROUTINE BALMINO
```

```
!--------------------------------------------------------
```

8) **Program Name** HANSENJF

Calling method: CALL HANSENJF(N, M, K, E, X, DX)

Function: Calculation of Hansen coefficients $X_k^{n,m}$ and their derivatives $\frac{dX_k^{n,m}}{de}$ using Eqs. (3.65) and (3.73)

Parameters: INPUT E eccentricity

N, M, K indexes

OUTPUT X $X_k^{n,m}$

DX $\frac{dX_k^{n,m}}{de}$

```
!-----------------------------------------------------------
      SUBROUTINE HANSENJF(N, M, K, E, X, DX)
      IMPLICIT NONE
      INTEGER N, M, K
      REAL*8 E, X, DX
      REAL*8 PI
      INTEGER NN
      REAL*8 WH, E2, X1, X2
      INTEGER I
      REAL*8 EE, MM, F, ROA, ROAN
      PI = 3.14159265358979323846264 3D0
      NN = 2000
      WH = PI / DBLE(NN)
      E2 = 1.0D0 - E * E
      X1 = 0.0D0
      X2 = 0.0D0
      DO I = 0, NN / 2 - 1
          EE = WH * (0.4226497308103742D0 + 2.0D0 * DBLE(I))
          MM = EE - E * DSIN(EE)
          F = EE + 2.0D0 * DATAN(E * DSIN(EE) / (1.0D0 + DSQRT(1.0D0 - E * E) - E * DCOS(EE)))
          ROA = 1.0D0 - E * DCOS(EE)
          ROAN = ROA**(N + 1)
          X1 = X1 + ROAN * DCOS(DBLE(M) * F - DBLE(K) * MM)
          X2 = X2 - ROAN * (DBLE(N) / ROA * DCOS(F) * DCOS(DBLE(M) * F - DBLE(K) * MM) &
              & + DBLE(M) / E2 * (2.0D0 + E * DCOS(F)) * DSIN(F) &
              & * DSIN(DBLE(M) * F - DBLE(K) * MM))
          EE = WH * (1.5773502691896257D0 + 2.0D0 * DBLE(I))
          MM = EE - E * DSIN(EE)
          F = EE + 2.0D0 * DATAN(E * DSIN(EE) / (1.0D0 + DSQRT(1.0D0 - E * E) - E * DCOS(EE)))
          ROA = 1.0D0 - E * DCOS(EE)
          ROAN = ROA**(N + 1)
          X1 = X1 + ROAN * DCOS(DBLE(M) * F - DBLE(K) * MM)
          X2 = X2 - ROAN * (DBLE(N) / ROA * DCOS(F) * DCOS(DBLE(M) * F - DBLE(K) * MM) &
              & + DBLE(M) / E2 * (2.0D0 + E * DCOS(F)) * DSIN(F) &
              & * DSIN(DBLE(M) * F - DBLE(K) * MM))
      ENDDO
      X = X1 * WH / PI
      DX =X2 * WH / PI
      RETURN
      END SUBROUTINE HANSENJF
```

D.3 Fortran Programs of the Calculation Methods in Chap. 4

```
!-----------------------------------------------------------
```

1) **Program Name** R6FWD

Calling method: CALL R6FWD(K, NN, E, X)

Function: Calculation of the initial values of the recursion using Eq. (3.17) or Eq. (3.87)

$$X_k^{n,m} \qquad (n = -l-1; m = -l, -l+1, \cdots, l; l = 2,3)$$

Then recursion calculation of eccentricity functions using Eqs. (R1), (R3) and (R6)

Parameters: INPUT K index
E eccentricity
NN order
OUTPUT X(N, M) X(3 : NN + 1, -NN : NN)

$$X_k^{n,m} \qquad (n = -l-1; m = -l, -l+1, \cdots, l; l = 2,3,\cdots,NN)$$

```
!--------------------------------------------------------------------
      SUBROUTINE R6FWD(K, NN, E, X)
      IMPLICIT NONE
      INTEGER K, NN
      REAL*8 E, X(3 : NN + 1, -NN : NN)
      REAL*8 E2, E1, E3, E4, E5, E6, E7, E8
      INTEGER N, M
      REAL*8 X1, A1, A2, A3, A4, A5, A6
      E2 = E * E
      E1 = 1.0D0 - E2
      E3 = DSQRT(E1)
      E4 = E3 * E1
      E5 = E1 * E1
      E6 = E2 / 4.0D0
      E7 = 1.0D0 - E2 / 2.0D0
      E8 = E1 * E
      DO N = 3, 4
          DO M = -N + 1, N - 1, 2
              CALL WNUK1(-N, M, K, E, X1)
              X(N, M) = X1
          ENDDO
      ENDDO
      DO N = 4, NN
          A1 = DBLE(N - 3) * DBLE(N + N - 1) * E2
          A2 = -3.0D0 * E2
          A3 = 2.0D0 * DBLE(N - 3) * DBLE(N - 2) + 3.0D0 * DBLE(N - 2) * E2 &
              & - 2.0D0 * DBLE(N - 3) * DBLE(K) * E4
          A4 = 3.0D0 * DBLE(N - 3) * DBLE(K) * E * E3
          A5 = DBLE(N - 5) * DBLE(K) * E * E3
          A6 = 2.0D0 * DBLE(N - 3) * DBLE(N - 2) + 3.0D0 * DBLE(N - 2) * E2 &
              & + 2.0D0 * DBLE(N - 3) * DBLE(K) * E4
          DO M = -N + 2, N - 2, 2
              IF(M .NE. 0) THEN
                  X(N + 1, M) = (DBLE(N - 1) * E * (X(N, M - 1) - X(N, M + 1)) &
                      & + 2.0D0 * DBLE(K) * E3 * X(N - 1, M)) / DBLE(M + M) / E1
              ELSE
                  X(N + 1, M) = (E8 * (X(N, M - 1) + X(N, M + 1)) + E7 * X(N - 1, M) &
                      & - E6 * (X(N - 1, M + 2) + X(N - 1, M - 2))) / E5
```

```
        ENDIF
    ENDDO
    X(N + 1, N) = (A2 * X(N + 1, N - 4) + A3 * X(N + 1, N - 2) &
        & + A4 * X(N, N - 1) + A5 * X(N, N - 3)) / A1
    X(N + 1, -N) = (A2 * X(N + 1, -N + 4) + A6 * X(N + 1, -N + 2) &
        & - A4 * X(N, -N + 1) - A5 * X(N, -N + 3)) / A1
ENDDO
RETURN
END SUBROUTINE R6FWD
```

!--

2) **Program Name** R5BACK

Calling method: CALL R5BACK(K, L, E, X)

Function: Calculation of the initial values of the recursion using Eq. (3.17) or Eq. (3.87)

$X_k^{n,m} \qquad (n=-l-1; m=-l,-l+1,\cdots,l; l=L)$

Then recursion calculation of eccentricity functions using Eq. (R5)

Parameters: INPUT K index

E eccentricity

L order

OUTPUT X(N, M) X(3 : L + 1, -L : L)

$X_k^{n,m} \qquad (n=-l-1; m=-l,-l+1,\cdots,l; l=L,L-1,\cdots,2)$

!--

```
SUBROUTINE R5BACK(K, L, E, X)
IMPLICIT NONE
INTEGER K, L
REAL*8 E, X(3 : L + 1, -L : L)
REAL*8 E1
INTEGER M
REAL*8 X1
INTEGER N
E1 = 2.0D0 * DBLE(K) * DSQRT(1.0D0 - E * E)
DO M = -L, L
    CALL WNUK1(-L - 1, M, K, E, X1)
    X(L + 1, M) = X1
ENDDO
DO N = L, 3, -1
    DO M = -N + 1, N - 1
        X(N, M) = (2.0D0 * DBLE(M) * X(N + 1, M) &
            & + DBLE(N + M) * E * X(N + 1, M + 1) &
            & + DBLE(M - N) * E * X(N + 1, M - 1)) / E1
    ENDDO
ENDDO
RETURN
END SUBROUTINE R5BACK
```

```
!-----------------------------------------------------------------
```

3）**Program Name** HANSENN

Calling method: CALL HANSENN(E, L, NN, Q0, G)

Function: Batch calculation of eccentricity functions and their derivatives of $l = L$ using Eqs. (3.65) and (3.72)

Parameters: INPUT

INPUT	E	eccentricity
	L	order
	NN	number of sections of numerical integration
	Q0	summation range of the index q $-Q_0 \sim Q_0$
OUTPUT	G(2, 0 : L, -Q0 : Q0)	
	G(1, P, Q),	$G_{Lpq}\left(p = 0, 1, \cdots, L; q = -Q_0, -Q_0 + 1, \cdots, Q_0\right)$
	G(2, P, Q),	$\dfrac{dG_{Lpq}}{de}\left(p = 0, 1, \cdots, L; q = -Q_0, -Q_0 + 1, \cdots, Q_0\right)$

```
!-----------------------------------------------------------------
      SUBROUTINE HANSENN(E, L, NN, Q0, G)
      IMPLICIT NONE
      REAL*8 E
      INTEGER L, NN, Q0
      REAL*8 G(2, 0 : L, -Q0 : Q0), PI, AA(2), E2, WH
      INTEGER P, Q
      REAL*8 WS(2, 0 : L, -Q0 : Q0)
      INTEGER I, J
      REAL*8 EE, M, F
      INTEGER N
      REAL*8 ROVERA, ROAN, FDE, GGC, GGS
      INTEGER MM, K
      REAL*8 S1, C1, WF1(2, 0 : L, -Q0 : Q0)
      PI = 3.141592653589793238462643D0
      AA(1) = 0.4226497308103742D0
      AA(2) = 1.5773502691896257D0
      E2 = 1.0D0 - E * E
      WH = PI / DBLE(NN)
      DO P = 0, L / 2
          DO Q = -Q0, Q0
              WS(1, P, Q) = 0.0D0
              WS(2, P, Q) = 0.0D0
          ENDDO
      ENDDO
      DO I = 0, NN / 2 - 1
          DO J = 1, 2
              EE = WH * (AA(J) + 2.0D0 * DBLE(I))
              M = EE - E * DSIN(EE)
              F = EE + 2.0D0 * DATAN(E * DSIN(EE) / (1.0D0 + DSQRT(1.0D0 - E * E) &
                  & - E * DCOS(EE)))
              N = DBLE(-L - 1)
              ROVERA = 1.0D0 - E * DCOS(EE)
```

```
          ROAN = ROVERA**(N + 1)
          FDE = (2.0D0 * DSIN(F) + E * DSIN(2.0D0 * F) / 2.0D0) / E2
          GGC = -ROAN * N * DCOS(F) / ROVERA
          GGS = -ROAN * FDE
          DO P = 0, L / 2
             MM = L - P - P
             DO Q = -Q0, Q0
                K = MM + Q
                S1 = DSIN(DBLE(MM) * F - DBLE(K) * M)
                C1 = DCOS(DBLE(MM) * F - DBLE(K) * M)
                WF1(1, P, Q) = ROAN * C1
                WF1(2, P, Q) = GGC * C1 + GGS * S1 * DBLE(MM)
             ENDDO
          ENDDO
          DO P = 0, L / 2
             DO Q = -Q0, Q0
                WS(1, P, Q) = WS(1, P, Q) + WF1(1, P, Q)
                WS(2, P, Q) = WS(2, P, Q) + WF1(2, P, Q)
             ENDDO
          ENDDO
       ENDDO
    ENDDO
    DO P = 0, L / 2
       DO Q = -Q0, Q0
          WS(1, P, Q) = WH * WS(1, P, Q) / PI
          WS(2, P, Q) = WH * WS(2, P, Q) / PI
       ENDDO
    ENDDO
    DO P = 0, L
       DO Q = -Q0, Q0
          IF(P .LT. L / 2) THEN
             G(1, P, Q) = WS(1, P, Q)
             G(2, P, Q) = WS(2, P, Q)
          ELSE
             G(1, P, Q) = WS(1, L - P, -Q)
             G(2, P, Q) = WS(2, L - P, -Q)
          ENDIF
       ENDDO
    ENDDO
    RETURN
    END SUBROUTINE HANSENN
```

References in Appendices A–D

1. Laskar J, Boué G (2010) Explicit expansion of the three-body disturbing function for arbitrary eccentricities and inclinations. Astron Astrophys 522:A60
2. Challe A, Laclaverie JJ (1969) Disturbing function and analytical solution of the problem of the motion of a satellite. Astron Astrophys 3:15–28
3. Jarnagin MP (1965) Expansions in elliptic motion. Astronomical papers prepared for the use of the american ephemeris and nautical almanac Volume XVIII, U.S. Government Printing Office, Washington
4. McClain WD (1978) A recursively formulated first-order semianalytic artificial satellite theory based on the generalized method of averaging. Volume II. The Explicit Development of the First-Order Averaged Equations of Motion for the Nonspherical Gravitational and Nonresonant Third-Body Perturbations. NASA CR-156783
5. Proulx RJ, McClain WD (1988) Series Representations and rational approximations for Hansen coefficients. J Guidance 11(4):313–319
6. Wu LD, Zhang MJ (2022) Direct calculation methods of Hansen coefficients and their derivatives. Chin Astron Astrophy 46(1):137–151
7. Giacaglia GEO (1976) A note on Hansen's coefficients in satellite theory. Celest Mech 14(4):515–523
8. Wu LD (2011) Orbit and detection of artificial satellites and space debris. China Science and Technology Press, Beijing

MIX
Papier aus verantwortungsvollen Quellen
Paper from responsible sources
FSC® C105338

If you have any concerns about our products,
you can contact us on
ProductSafety@springernature.com

In case Publisher is established outside the EU,
the EU authorized representative is:
Springer Nature Customer Service Center GmbH
Europaplatz 3, 69115 Heidelberg, Germany

Printed by Libri Plureos GmbH
in Hamburg, Germany